wild
collections

This book is dedicated to my friends and colleagues at CSIRO, whose dedication to science is creating a healthier natural environment and a brighter future for us all.

wild collections

SPECIMENS, STORIES AND SCIENCE from **CSIRO**

Andrea Wild

© Text CSIRO 2025

All rights reserved. Except under the conditions described in the *Australian Copyright Act 1968* and subsequent amendments, no part of this publication may be reproduced, stored in a retrieval system or transmitted in any form or by any means, electronic, mechanical, photocopying, recording, duplicating or otherwise, without the prior permission of the copyright owner. Contact CSIRO Publishing for all permission requests.

Andrea Wild asserts their right to be known as the author of this work.

 A catalogue record for this book is available from the National Library of Australia

ISBN: 9781486318810 (pbk)
ISBN: 9781486318827 (epdf)
ISBN: 9781486318834 (epub)

Published by:

CSIRO Publishing
36 Gardiner Road, Clayton VIC 3168
Private Bag 10, Clayton South VIC 3169
Australia

Telephone: +61 3 9545 8400
Email: publishing.sales@csiro.au
Website: www.publish.csiro.au
Sign up to our email alerts: publish.csiro.au/earlyalert

Front and back covers: Detail of a *Lophorina* bird-of-paradise in the Australian National Wildlife Collection (photo by Martin Ollman, © CSIRO)
Back cover: Scarab beetle in the Australian National Insect Collection (photo by Cate Lemann, CSIRO)

Edited by Adrienne de Kretser, Righting Writing
Cover design by Cath Pirret
Typeset by Envisage Information Technology
Index by Indexicana

CSIRO Publishing publishes and distributes scientific, technical and health science books, magazines and journals from Australia to a worldwide audience and conducts these activities autonomously from the research activities of the Commonwealth Scientific and Industrial Research Organisation (CSIRO). The views expressed in this publication are those of the author(s) and do not necessarily represent those of, and should not be attributed to, the publisher or CSIRO. The copyright owner shall not be liable for technical or other errors or omissions contained herein. The reader/user accepts all risks and responsibility for losses, damages, costs and other consequences resulting directly or indirectly from using this information.

CSIRO acknowledges the Traditional Owners of the lands that we live and work on across Australia and pays its respect to Elders past and present. CSIRO recognises that Aboriginal and Torres Strait Islander peoples have made and will continue to make extraordinary contributions to all aspects of Australian life including culture, economy and science. CSIRO is committed to reconciliation and demonstrating respect for Indigenous knowledge and science. The use of Western science in this publication should not be interpreted as diminishing the knowledge of plants, animals and environment from Indigenous ecological knowledge systems.

Apr25_RP_ILS

FOREWORD

I grew up and then raised my children by the Yarra River (Birrarung) in an outer suburb of Melbourne called Warrandyte that was, and to some extent has remained, an island of bush in a sea of expanding urban sprawl. That experience has sheeted home to me what has been obvious to First Nations people for millennia, that there is an inextricable link between the health of the environment and human flourishing.

School and university were a mixed experience for me: I have always been curious, enjoyed learning, and for as long as I can remember have been fascinated by science, but struggled to balance obsession with certain topics with the discipline of studying the curriculum evenly. It wasn't until I spent a summer in a laboratory at the John Curtin School of Medical Research in ANU that I realised there was a job – a research scientist – that paid you to curiously obsess. For the first time in my life, I felt like I didn't have to explain myself, I found my community and fell in love with blood cells. Studying how blood cells communicate has been my professional life for 40 years.

Loving work makes holidays difficult. Within 24 hours I'd be bored, sitting under a beach umbrella itching to get back to the lab, dreaming about experiments I wanted to do. The answer was holiday science and since I spent vacations in national parks where biodiversity is incredible, I developed an interest in entomology and very quickly, an obsession with tiny moths.

Just as I found a community in medical research institutes working on blood cells, I have been privileged to find an incredibly welcoming community at CSIRO's Australian National Insect Collection – Dr Marianne Horak and the late Ted Edwards and Dr Ian Common were not just encouraging and supportive, but became friends and mentors. Over the last decade, with a group of other passionate citizen scientists, I've studied heliozelids, the sun-loving moths. These tiny, silvery moths are active during the day, like butterflies, and pollinate Australian flowering plants such as boronias. You can read about them in Chapter 10 of this book.

Spending time in the insect collection revealed to me how valuable natural history collections are as libraries of life. Biodiversity specimens tell us what exists in the world and where. They inform how to monitor and manage the environment, how to recover endangered species, how to deal with pest insects and weeds, how to prevent diseases and pandemics, and how to use the natural world to sustain our lives. Above all, our collections reveal the interconnectedness of nature. They inspire an awe and respect for the extraordinary biological diversity of our planet.

When I think about how a big, complicated organisation like CSIRO delivers impact from its collections, the answer is clearly its people. If you follow a tiny moth from a boronia flower into the insect collection and through the research that unveils the secrets of the natural world, you will find a network of people collaborating on things that matter to all of us.

Our people have hundreds of stories to tell you.

CSIRO is a magical place. In this inspiring book you'll discover why.

Dr Doug Hilton AO, Chief Executive, CSIRO

CONTENTS

Foreword		v
Preface		x
Acknowledgements		xi
Introduction		xii
1	**Eye of newt and toe of frog**	**1**
	Blind-worm's sting	1
	Owlet's wing	3
	Come, you spirits	8
	Wool of bat	11
	A moving wood	13
2	**An alien wasp from Jupiter**	**15**
	An alien among aliens	15
	A fish out of water	17
	An orchid, underground	18
	A parrot in the night	21
	A sponge filled with sharks	24
	A fly in a toilet	27
3	**Marvellous names**	**31**
	What's in a name?	31
	Choosing a name	33
	Distant cousins in the sea	36
	Such galahs	40
	Stan Lee and his insect assassins	42
	Un-naming	43

4	**Cocktails in nature**	**47**
	Cocktails of spider wasps	47
	The language of lichens	49
	Diving into algae	52
	A living collection	54
	Fossil eyes and plastic munching	58
5	**Lost in time**	**61**
	Paradise lost	61
	Extinct insects	65
	Neither here nor there	68
	Living giants of the deep	70
	Giant birds and a lost shark	72
6	**Curiouser and curiouser**	**77**
	Dung rollers	77
	Mermaids' purses	80
	Resident weevil	83
	Jumping fossils	86
	Deep sea strangers	89
7	**Space invaders**	**93**
	Tree body problem	93
	An ancient problem of biblical proportions	94
	Parrots of the world	97
	Diving with handfish	100
	Weeds, bugs and keys	102
	An endangered weed	106
8	**Tales of genitalia ... and other strange ways to get information about animals**	**109**
	Quirky tails	109
	More tales of tail ends	111

	The love lives of sharks	113
	Wild tales of reproduction	116
	Buckets of gunk	118
	Out of thin air	121
	Jurassic lark	125
9	**Big moves and quantum leaps**	**129**
	A new home	129
	Pinning down the move	133
	A decathlon of beetles	137
10	**A species on life support**	**143**
	Why does biodiversity matter?	143
	Beauty and the bees	146
	Tiny moth pollinators	150
	A significant skink	153
	Saving an ocean forest	154
	Glossary	160
	Plates	163
	Index	179

PREFACE

I'm a science communicator at CSIRO. I've spent my career reading papers, writing stories and working alongside great scientists.

Since 2016, I've been working with the National Research Collections Australia, supporting the scientists to appear in the media and promote their research to the public. Every day I'm overwhelmed by their dedication to science for the environment and overwhelmed by their generosity and kindness. But even more, I'm amazed by the stories they tell me, the specimens they show me and the astonishing titbits they drop into conversations.

In this book, I invite you to join me as I explore CSIRO's collections and meet the people working toward a brighter future for Australia's biodiversity.

Andrea Wild

ACKNOWLEDGEMENTS

This book was written on the lands of the Ngunnawal People and we pay respect to their Elders past and present. We acknowledge the Traditional Owners of the many lands across Australia from where the specimens in the National Research Collections Australia have been collected.

Thank you to my colleagues at CSIRO's National Research Collections Australia for sharing stories about their research and reviewing sections of this book. Particular thanks to Helen O'Neill, Leo Joseph and Lucinda Ross.

Thank you to my editors Melinda Chandler (who suggested I write this book) and Mark Hamilton for their feedback to shape this book.

Finally, thank you to the collectors of the past who have made the National Research Collections Australia the amazing resource they are today.

INTRODUCTION

The National Research Collections Australia at CSIRO are natural history collections. Unlike many other preserved biological collections, these are not museums. Instead, our collections are a resource for research and an important part of Australia's research infrastructure. They include:

- Australian National Algae Culture Collection
- Australian National Fish Collection
- Australian National Herbarium*
- Australian National Insect Collection
- Australian National Wildlife Collection
- Australian Tree Seed Centre
- Australian Tropical Herbarium**

We use our collections to better comprehend Australia's largely unique biodiversity, its origins and relationships, and how it is changing. The better we understand our biodiversity, the more we can apply this knowledge to the natural world, helping others. Active examples are the prevention of weed incursions, improved monitoring of threatened species in ecosystems, and the current work underway restoring Tasmania's Giant Kelp forests.

The collections hold more than 15 million specimens, many collected overseas through bilateral agreements and

---

* The Australian National Herbarium is a joint venture between Parks Australia, which includes the Australian National Botanic Gardens, and CSIRO.
** The Australian Tropical Herbarium is a joint venture between James Cook University, CSIRO, Parks Australia and the Queensland Government.

permits as part of collaborative projects, or donated through specimen exchanges.

We are the custodians of plant and animal specimens collected from the traditional lands of many Aboriginal and Torres Strait Islander Peoples.

As the director of CSIRO's National Research Collections Australia, I am working with our collections to recognise the contributions, interests and rights of Aboriginal and Torres Strait Islander Peoples. We are beginning to work with Traditional Owners to make the collections more accessible to people working on Country, include Indigenous perspectives in scientific research and ensure benefits are shared. We look forward to working in partnership.

Anthony Whalen,
Director, National Research Collections Australia, September 2024

1
Eye of newt and toe of frog

It's the winter of 2022 and a neurosurgeon in Canberra is operating on a person's brain. The patient has been experiencing a mystery illness and scans have revealed a mass in their right frontal lobe. As the operation progresses, the neurosurgeon finds the cause. It's a bright red, 8 cm worm which is now wriggling at the end of her forceps.

Blind-worm's sting

The patient was the first known human to be infected by *Ophidascaris robertsi*, a parasitic nematode normally at home in carpet pythons.

Nematodes are so plentiful in nature that if they were lined up nose to tail they would stretch light years across space. Many are microscopic but some can be several metres long. The record is an 8.4 m parasite found in the placenta of a sperm whale. Nematodes have been found almost everywhere on Earth, from deep inside gold mines to within the ears of reindeer.

Many species of parasitic nematodes and other kinds of parasitic worms live in the guts of animals, such as in the liver or intestines. Some spend time in other parts of the body too, including the lungs, muscles, blood, brain or beyond. Some have extraordinary life cycles that require multiple host species or involve time spent living freely in the environment. Some can cause zoonotic infections in humans.

The worm removed from the person's brain was identified by Dr Dave Spratt, an honorary parasitologist at the Australian National Wildlife Collection. Together, we look at a vial of ethanol containing a preserved fragment of this worm.

Dave tells me *O. robertsi* is a parasite of carpet pythons but requires an intermediate host to complete its life cycle. Infections can also spill over into accidental hosts. Adult worms live in the carpet python's gastrointestinal tract and their eggs are shed in its faeces. The eggs are sticky and can remain on vegetation that is later eaten by animals.

'Inside the egg, the larva develops to become a second-stage larva, or L2,' Dave says. 'If the egg is eaten by an intermediate host, such as a small marsupial, it triggers the L2 to hatch. The L2 migrates to the liver and matures into a third-stage larva, or L3, which is around 8 cm long. If a carpet python then catches and eats that animal, the L3 matures into an adult in the snake's gut and the life cycle begins again.'

Dave explains that *O. robertsi* is not fussy about its intermediate host. It can also infect accidental hosts – animals that aren't normally on the menu of carpet pythons. Accidental hosts are a dead end for the parasite in terms of its life cycle, but the L3 are very active. They can leave the liver and wander through tissues and organs, reaching distant areas like the brain.

'This nematode has been recorded in many kinds of native and non-native mammals, including koalas and possums. An animal can be infected by dozens of larvae,' Dave says. 'I've seen this parasite so many times in wildlife that as soon as the surgeon sent me a photo of the living worm, I knew from its bright red colour what it was.'

Dave's diagnosis was later confirmed by DNA work. After a lifetime of working with the parasites of Australian wildlife, it was not the first time he had seen nematodes cause cryptic diseases in humans. One that has him particularly concerned is *Haycocknema perplexum*, the muscle-destroying nematode.

'*H. perplexum* is microscopic and lives inside muscle cells. The eggs hatch inside their mother and then develop in her to third-stage infective larvae. They then burst from her head, creating a cascading infection in muscle cells that can be fatal,' Dave says.

Dave named the species and chose the name '*perplexum*' because the species is perplexing. No one knows the details of its life cycle, how people or wildlife get infected or whether it can survive outside its host in the environment.

Fewer than 10 infections of this nematode have been diagnosed in humans in Australia. It's thanks to Dave and a medical colleague that there is now a successful treatment for people with *H. perplexum* infections: the same drug used in 'sheep drench' – a treatment safe for humans but better known for treating nematode infections in livestock.

Owlet's wing

In Shakespeare's *Macbeth*, the grisly ingredients tossed into the witches' cauldron may have been plants, named by herbalists after the animal features they resembled. Newts' eyes were mustard seeds. Frogs' toes were the bulbous stems of buttercups. Owlet's wing was also a plant, perhaps garlic or perhaps the spicy root of ginger, one of many species in the family Zingiberaceae.

I join a group of trainees from the Australian National Herbarium to learn the techniques of botanical collecting.

We'll be searching for an ornamental ginger on the escarpment between Canberra and the beaches a couple of hours' drive north-east. Our four-wheel drive and van are filled with the low-tech equipment of botanical collectors: secateurs, old newspapers, cardboard, timber presses and rope.

Brendan Lepschi, head curator at the herbarium, is keen for us to collect some specimens of Ginger Lily (*Hedychium gardnerianum*) to fill a gap in the herbarium's collection. Native to the Himalayas, this attractive plant has been introduced to Australia as a garden plant and often escapes the borders of people's yards. We soon spot its tall yellow and orange flowers and dark green leaves on the side of the road. The grass is tall and lush and lichens cover the trunks of nearby trees. It's pouring rain and I have forgotten to bring a coat.

The trainees use secateurs to take specimens of the ginger plants, slicing through their thick green stems. They pencil notes about the specimens in a field notebook: what species is it, what are its features, where are we, who collected it, what else is growing nearby, what's the soil like? We stop several times to collect specimens from different Ginger Lilies growing in different locations.

'Multiple specimens are important for herbaria,' Brendan says. 'They show the variation that exists within a plant species and how it differs among locations or changes over long stretches of time. They also serve as a record of spatial occurrence.'

Eventually we arrive in a small town with a picnic ground scattered with tall eucalypts. We unpack the specimens and huddle around a picnic table that has a hut-like roof, providing a small amount of shelter from the rain. Here we begin the preservation process.

1 – Eye of newt and toe of frog

Brendan picks up a specimen, removes a sample of its flowers and slides it inside a vial of ethanol. This will preserve the three-dimensional arrangement of the petals and the way the individual flowers spring from the stem.

One of the trainees then takes the remainder of the specimen, cuts its stem in half to conform to the standard size of a herbarium specimen and arranges it carefully between sheets of newspaper, labelling the pieces to match the record in the field notebook. Many hands repeat the steps with the other specimens and then they are sandwiched between sheets of corrugated cardboard inside timber press boards cinched tightly with rope.

The Ginger Lily specimens have a high moisture content and are wet from the rain. If this were a longer field trip, they would be removed from the presses each day and their papers and cardboard changed to prevent the specimens rotting or growing mould. Brendan tells me that changing papers by torch light at a camp site is a typical evening for a botanical collector on a field trip.

Back in Canberra, the specimens enter the herbarium's receivals room and are placed in a drying oven at 40°C. Lush, thick specimens like this can take several weeks to fully dry. Their papers need to be changed every few days. Sometimes plants like these need their stems to be sliced open so they dry more quickly, without turning mouldy.

'Cacti are difficult. You need to fillet them like a fish and salt them to draw out the moisture,' Brendan tells me, laughing.

When the specimens are completely dry, they will spend two weeks in a freezer at −18°C to kill any pest insects or insect eggs present on the plants. The kinds of creatures that clean up dead plants in nature would happily feast on the 1+ million plant specimens in the herbarium.

I visit the herbarium's mounting room to witness the final step in preparing a specimen. The art of mounting specimens is carried out mainly by volunteers. They place a large sheet of archival card in portrait orientation and lay the specimen with its roots or stems at the base of the sheet and its flowers and leaves at the top, attach the specimen with archival thread or tape, and in the lower right corner glue a label containing information transferred from the field notebook. Next, herbarium specimens are digitised so that a high-resolution photograph and databased information from the label can be shared online with the world.

Some specimens end up resembling works of art (Plate 1). Others are simply what they are: an invaluable and irreplaceable resource for research. After specimens are pressed, dried, mounted and identified, they are sorted by species and geography and stored on mobile shelving units. There they wait to be studied, loaned by other researchers in Australia or overseas or sampled for DNA.

The aisles of mobile shelving can be opened and closed using steering wheels that remind me of bumper cars. A poster warns that certain specimens, like daisies and legumes and anything with protein-rich seeds, are vulnerable to hungry insects, which will also eat mould growing on the specimens themselves or old adhesive and tape. Carpet beetles and book lice are the top enemies.

'We don't use any chemicals here, we just move specimens to the freezer for a couple of weeks if we notice any insects on them,' Brendan says. 'Kept dry and safe from insect attack, pressed plants will last indefinitely.'

'The oldest specimens we have here are more than 250 years old. They were collected by Banks and Solander at Endeavour River in 1770 and held in the British Museum for around 200 years before they were sent here to us.'

One of Banks and Solander's specimens is a white gum, *Eucalyptus platyphylla*. Two little twigs with leaves and buds are glued to a sheet of paper. In life its leaves were a soft green, but as a dried specimen they are brown and brittle. An envelope stuck to the lower left corner of the sheet holds fragments of leaves that have broken off the little twigs. There are multiple labels, recording the species name, when and where it was collected and by whom. Other labels and stamps identify the herbarium and the specimen's accession number.

'Some sheets end up with multiple labels when people add notes or revise the classification and name of a specimen,' Brendan says.

The Ginger Lily flowers Brendan placed in ethanol during our collecting trip don't need any further treatment to preserve them. They are ready to enter the spirit collection. The ethanol, like hand sanitiser, will prevent any growth of bacteria or mould. Over time, the flower's yellow petals and orange stamen will leach their colours into the fluid until ghostly, whitish flowers remain floating in the vial. Curators will top up the liquid as it slowly evaporates. Scientists will take the vial off the shelf, peer through its glass, dissect a flower, and return it to the shelf. Over time it may be used to distinguish different species, describe plants that are new to science, report the distribution and spread of weeds, create apps that tell species apart, observe how flowers respond to climate change and myriad uses yet to be imagined.

Come, you spirits

At the foot of Black Mountain in the centre of Canberra, the grassy woodland gives way to a field of research laboratories. The newest building is Diversity, home to more than 12 million specimens of wildlife, insects and plants.

In the middle of the building is a cool, concrete bunker – the spirit vault. The name suggests a stash of whiskies or something supernatural, but the spirit vault is an assortment of plant and animal specimens stored in ethanol. Thousands of vials, jars and buckets hold a menagerie of specimens, from the macabre to the sublime.

Dr Clare Holleley leads the vertebrate collections at CSIRO, spanning fish and wildlife. Together we walk around the spirit collection, passing along a corridor of shelves lined with jars containing reptiles. One jar holds an army of frogs with blue-coloured backs and light-coloured bellies. In another, a lounge of small lizards is a tangle of bodies and tails. Some of the smallest specimens are in the several hundred vials preserving the tongues of birds. Most are long and thin. Some are textured with barbs and ridges, allowing each species to specialise in its diet, from nectar to fish.

Clare notices a jar of rats containing pieces of Christmas paper. Each rat has a shiny paper tag tied to one of its rear legs, revealing what species it is and when and where it was collected.

'The collectors must have run out of specimen labels and needed to improvise,' she says.

Clare tells me the method used to preserve vertebrate specimens this way locks up the three-dimensional structure of their DNA. This is because the specimens spend several weeks in formaldehyde before long-term storage in ethanol. Clare's team recently discovered that, with the right molecular approach, it is possible to tease out what the DNA was doing over a century ago.

'We can take a specimen and look at what genes were switched on at the time it died,' she says. 'We can compare the genes switched on and off in specimens collected at different times or in different places. This lets us ask questions about the past. What genes are important for animals experiencing environmental stress? How do they survive during times of drought? No one's ever been able to do this before.'

We turn a corner and leave the vertebrates behind. In front of us the flowers of 17,000 orchids float in small vials and jars of ethanol (Plate 2). Many of the vials have been dyed shades of orange, green and blue as the petals slowly leach their colours into the liquid. Stored in this way, as an alternative to being dried and pressed, the flowers retain their three-dimensional structure. They are a record of the unique features of Australia's estimated 1,600 orchid species.

Nearby, a bunch of wild bananas in a jar have turned a dark amber colour. They are small and slim, like a monkey's fingers, with seeds visible along their lengths. They were collected in 1964 in north-eastern Papua New Guinea (PNG). The field notebook from the trip reveals in the botanical collector's neat printing that the banana tree was 15 m tall. It was growing among *Castanopsis* trees in a secondary forest – one recovering following human disturbance – at around 1,300 m above sea level.

The largest floral specimen in the spirit collection is a flower that looks like a large red cylinder bursting with petals.

It's a *Tapeinochilos*, collected in 1999 growing on the Rimba Irian Golf Course in Papua. The collector noted that undisturbed lowland tropical rainforest was growing within and surrounding the greens of the golf course.

Herbarium staff, including Peter Gray, recently spent many months sorting out the plants stored in spirit before they were moved into this new spirit vault. Peter has kept a selection of old specimen jars to display during tours. They illustrate the challenges encountered during the move.

'There were antique bottles with glass stoppers and old jam jars still with their original labels,' Peter says. 'Some old jars had their lids rusted in place, which we needed to cut open to rescue the specimens inside.'

The curators also replaced the liquid of any specimens stored in formaldehyde. The new spirit mix is basically hand sanitiser: 70 per cent ethanol, 20 per cent water and 10 per cent glycerol. It supports the structure of the floral specimens, prevents microbes attacking them and is safe to handle. They call it BANG mix, an anagram formed from the initials of the Australian National Botanic Gardens (ANBG), whose staff perfected the recipe.

The last thing to see in the spirit vault is the helminth collection, where hundreds of tiny glass vials hold parasitic worms as a record of the biodiversity of our faunal biodiversity. Pick up a couple of vials at random and you might be holding noodle-like nematodes from a sea lion or the long, flat bodies of tapeworms from the gut of a koala. The fragment of *O. robertsi*, the worm removed from the person's brain, is here. Not far from it is a tall glass cylinder of ethanol containing the coils of its normal host, a carpet python.

Wool of bat

The witches of *Macbeth* also tossed an ingredient called wool of bat into their cauldron. It may have been moss, a cryptogam.

The Australian National Botanic Gardens next door to CSIRO in Canberra are home to the Australian National Herbarium's cryptogam collection. I tag along while curator Chris Cargill shows a group of CSIRO staff around the collection's cabinets and shelves.

'Cryptogams include plants like mosses, hornworts and ferns, that reproduce by spores instead of seeds,' Chris says. 'We also have lichens, slime moulds, seaweeds and fungi here.'

She slides a box from a shelf to reveal the largest of the 12,000 fungi in the collection, a brown-coloured bracket fungus in a clear plastic bag. It's 40 cm wide with an off-centre hole where it grew around a sapling. The surface of the fungus resembles the growth rings of a tree. Its label is blue, indicating it was collected outside Australia, in this case PNG.

'We fumigate the collection here once per year,' Chris tells us. 'Insects like to eat dried fungi, but they tend to leave specimens like mosses and hornworts alone.'

I ask Chris to show us a slime mould, imagining a bright orange jelly creeping along a forest floor. She pulls out a small cardboard box lined with a piece of white card that she says is archival and acid-free. On the card are glued four small pieces of bark, each with a bristle that looks like burnt feathers.

'These bristles are the fruiting bodies of a slime mould,' Chris explains.

There's handwriting all over the card. I note the accession number, HL99041, the species name, *Stemonitis flavogenita*, and a curious comment: 'Spores – very delicately warted'.

One of the people on the tour asks Chris whether there are any living plants in the collection.

'I have a glasshouse with a few plants like hornworts growing,' Chris says. 'I bring them back from the field to allow the fruiting part of the plants that produce the spores to mature, as many of the species I study have ornate spores important for their identification.'

Hornworts are small, not very common plants with horn-like structures, hence their name. Wort simply means plant, particularly one associated with medicinal properties. There are fewer than 250 known species of hornworts worldwide. Over the past few years, Chris has scientifically named four new Australian species, bringing the total number of Australian hornworts to 27.

The recently named hornworts belong to a genus called *Anthoceros* (Plate 3). They are frilly looking, with internal cavities filled with mucilage. To tell the species apart, you need to compare the patterns on their microscopic spores by taking scanning electron micrographs. These tiny but crucial details are completely invisible to the naked eye.

Aside from the hornworts and a few other plants Chris keeps in the glasshouse, only dead specimens are allowed inside the herbarium. But death is a relative term here. Even after many years stored in collections, some species of moss can come back to life if they are soaked in water. One of the mosses held here is *Chorisodontium aciphyllum*, a species collected from Antarctica. It forms huge green beds that grow incredibly slowly, just a few millimetres per year. It can survive freezing, not just for a season but for hundreds of years. As glaciers melt elsewhere in the world, people are discovering other species of moss still alive after being frozen for hundreds of years.

A team from the imaging company Picturae is in the middle of digitising the cryptogam collection. Today they are photographing moss specimens. At one end of an 8 m long conveyor belt set up in the middle of the herbarium, a person unpacks bright blue paper packets holding little plastic bags of dried moss specimens that were collected all over the world. A second person straightens the specimens and their packets before they pass into a cabinet where a camera takes a high-resolution photograph. A third person packs the specimen away. It will take the team almost a year to work through the 300,000 specimens in the cryptogam collection.

Brendan tells me people began photographing specimens at the herbarium back in the 1940s, showing the desire to duplicate and share collections has been around for a long time. Only recently, though, have we had the digital technology to achieve this at scale.

'People here started taking specimen photos with Box Brownie cameras,' he says. 'Those black and white images are tiny and not very useful, but the idea was there.'

A moving wood

Walking back to the CSIRO site, I pass a tiny forest of eucalypts growing in large potting bags. It's a seed orchard of the threatened species Camden White Gum (*Eucalyptus benthamii*).

The tall, white trunks of these majestic trees were once a common sight along the Nepean River south-west of Sydney. Today, the remnant trees are growing too far apart to pollinate each other. This means they don't set high-quality seed for the next generation.

Dr David Bush, director of the Australian Tree Seed Centre, says the seed orchard at CSIRO is part of the answer.

'We're growing clones – exact genetic copies of the trees in the wild,' he says. 'We grafted small cuttings of the most threatened wild trees onto rootstocks. We now have several dozen clones that flower, pollinate each other and set seed.'

'We are doing this work in Canberra so we can ensure the genetic purity of the seed. But our intention is to plant the resultant seedlings back along the Nepean.'

'Once the seedlings have grown, they are being planted by Camden Council and other community groups in the Camden area. When they mature, they will pollinate the remnant wild trees, helping conserve the species.'

Many Australian hardwoods are prized overseas for pulp fibre, as construction materials and as fuel for heating, cooking and industrial processes. As well as taking part in conservation projects, the tree seed centre supplies seed to forestry plantations around the world. Its seed vault has row after row of large silver canisters filled with tree seeds.

'Camden White Gum is a subtropical tree that is also frost-tolerant,' David says. 'This means it is suitable for plantations in some regions of Brazil, Uruguay and the USA that have generally warm weather but can get sudden cold spells. It's one of the most widely planted eucalypts in the world, despite being vulnerable in its natural setting.'

> *Next, I decide to visit the Australian National Insect Collection, where something even more surprising than a brain worm awaits.*

2
An alien wasp from Jupiter

In 1920, the entomologist A.A. Girault wrote a scientific paper titled 'Some Insects never before seen by Mankind.' In this paper he described four tiny species of wasps. Three were from Queensland, including one conveniently discovered on a window in an entomology lab. The fourth was found much further away: supposedly in a chasm on Jupiter.

An alien among aliens

When I first meet Dr Ian Naumann, a retired research fellow at the Australian National Insect Collection, he is in his office looking down a microscope at a tray of tiny creatures. He tells me they are tea leaves. From where I'm standing in the doorway, this could be true. The flecks glued to tiny cardboard tags pinned in the tray do indeed look like tiny tea leaves rescued from the bottom of a cup, but their tiny legs give the game away.

The tea leaves are actually parasitoid wasps, the creatures that were an inspiration for the Xenomorph in the *Alien* movies. It turns out that bursting from the body of your host is not only real, in the insect world it's common.

Female parasitoid wasps lay their eggs in the eggs and bodies of other insects or spiders. The larval wasps hatch, eat their hosts from the inside out, pupate and emerge as adults.

There are thousands of different species. They range from tiny wasps that parasitise the eggs of other insects, to giant wasps that paralyse large spiders with stings that are painful to humans. The venomous cocktails of spider wasps keep spiders paralysed and fresh for long periods. They're being explored for new pharmaceuticals (see Chapter 4).

Ian's tea leaves belong to the genus *Anastatus* in the family Eupelmidae. Wasps in this genus parasitise the eggs of plant-sucking bugs and praying mantids and the cocoons of beetles. Ian tells me parasitic wasps tend to lay one egg per victim.

'Parasitoid wasps are generally pretty good at laying the right number of eggs per host to ensure there is enough food in the parcel to feed the offspring,' he says. 'Most species are very small because the hosts they feed on are small.'

'Eupelmids do most of their feeding as larvae. Adults may visit flowers for a sugar meal and females of some species bite hosts for a protein meal to help them produce eggs.'

In nature, parasitoid wasps are so common they keep insect numbers under control. They can also be used for biocontrol. In fact, researchers hunt for parasitoid wasps that target invasive species. Ian says some of the eupelmid species he studies might target pest stink bugs, such as the Brown Marmorated Stink Bug (see Chapter 7). Preventing this crop pest from establishing in Australia is a top priority, but using parasitoid wasps to attack the bugs could be one of the answers if it does turn up.

Ian notes the job of describing Australia's parasitoid wasp species has not been made easier by A.A. Girault, who described a shadowless, airy wasp species collected from a naked chasm on Jupiter on 5 August 1919. He wrote that it was 'visible only from certain points of view' and that its flight could not be followed 'except by the winged mind'.

Girault was a self-styled entomologist who wrote papers with ethereal titles like 'Some Gem-like or Marvellous

Inhabitants of the Woodlands Heretofore Unknown and by Most Never Seen nor Dreamt of'.

Ian tells me Girault self-published many of his research papers without peer review, the process where a scientific journal has papers checked by scientists who are experts in the field.

'Girault described many species that he or his colleagues had collected, but he also described things that don't exist from places that don't exist,' Ian says. 'He seems to have based the description of the wasp from Jupiter on an entomologist whom he didn't like.'

A fish out of water

When Lake Pedder in Tasmania was dammed in 1972, the waters of the lake were opened to invasive Brown Trout (*Salmo trutta*) and a native galaxiid species that displaced its rarer relative, the Pedder Galaxias (*Galaxias pedderensis*). Today, the latter species is one of Australia's most endangered fish.

Dr Bruce Deagle is a geneticist at the Australian National Fish Collection who works on Tasmania's galaxiids. I catch up with him to learn more about these small freshwater fishes.

'The Pedder Galaxias is considered extinct in the wild because it no longer exists in its original location,' Bruce tells me. 'But this species does actually survive in two other lakes, where it was transplanted to ensure its survival.'

The related Swan Galaxias (*Galaxias fontanus*) is a slightly smaller Tasmanian galaxiid species that grows up to 13.5 cm. It's mottled olive on top and silver below.

'This species is also highly endangered and survives only in small sections of a few streams,' Bruce says. 'We don't know its original distribution.'

'The streams where it survives today are above waterfalls, where Brown Trout can't colonise. But Swan Galaxias was described as a species about a century after Brown Trout were introduced to Tasmania, meaning its range was likely much bigger before it was displaced by Brown Trout.'

Bruce tells me the streams where Swan Galaxias live today are tiny, only a metre or two wide, and they are confined to small areas generally a kilometre or so in length. As a result, Swan Galaxias are quite vulnerable to environmental change.

'We recently conducted a population genomics study of Swan Galaxias to help decide on sampling to create new populations,' he says. 'The goal is to select individuals that represent the full genetic diversity of the species.'

The alternative is to take a random sample and hope for the best. I imagine this is a little like grabbing a handful of jellybeans from a jar and hoping that you have one of each colour. Population genetics enables conservationists to make decisions informed by science.

An orchid, underground

The total number of native species in Australia is more than 500,000. This is an estimated tally of animals, plants, fungi and algae, some of them microscopic. Around three-quarters of these are endemic to Australia, meaning they don't occur anywhere else on Earth.

'What is it?' and 'Where does it live?' are questions we still can't answer for two-thirds of Australian species. Only around one-third have been scientifically named and described. The rest, including many of Australia's unique orchids, are invisible to science.

Dr Heidi Zimmer is a research scientist at the Australian National Herbarium. After studying the famous Wollemi Pine, she turned to orchids, trained by orchid expert Dr Mark Clements. Heidi is standing in the office they share, surrounded by books, journals and filing cabinets overflowing with floral cards. These specimens are orchid flowers, dissected and taped to small rectangles of cardboard to record the exact shape and size of the flower parts of different species.

'Our goal is for no Australian orchid to go extinct,' Heidi tells me. 'There are more than 1,500 species and counting. We're still finding orchids in the wild, scientifically naming them and assessing whether they are threatened with extinction.'

If you try to picture an orchid, you might imagine a pot plant with a long stem of pink or white flowers springing from a base of smooth, dark green leaves. If so, you're thinking of a variety of *Phalaenopsis*, or moth orchid, common in the horticulture trade.

In the wild, most moth orchids are epiphytic – they grow on trees, like most of the world's orchids. In contrast, the majority of Australia's native orchids are terrestrial – they grow on the ground. Species of *Rhizanthella*, the underground orchid, take this to the extreme.

'The idea of a subterranean orchid is like life on Mars,' Mark tells me. 'They are so rare, I never expected to see one.'

Rhizanthella first caught the attention of scientists in 1928. In the century since, they've been found at only a handful of sites scattered across the country. There are now five scientifically recognised species and all are at risk of extinction. These orchids spend their whole lives underground and even flower beneath the soil's surface. They have a small, pale rhizome (a rootstalk or root-bearing underground stem) that develops a single pinkish-red flower around the size of a 50-cent piece. Its fruits are like tiny ball bearings.

Dr Katharina Nargar researches orchids at the Australian Tropical Herbarium in Cairns (Plate 4). She says these fruits are highly unusual for orchids.

'Orchids normally have capsules with tiny dust-like seeds which are dispersed by wind,' Katharina says. 'The underground orchid has little berry fruits, which points to animal-assisted seed dispersal. This is an adaptation of these orchids to their life underground.'

'Underground orchids have a very peculiar distribution. There are only a small number of populations on either side of the continent, in south-east and south-west Australia.'

Conservation of these orchids is complicated. They are killed by bushfires and burning. Their seeds need a particular fungus to germinate. They are very difficult to grow in pots. They depend on unknown animals to pollinate their flowers and disperse their seeds. If these animals are locally extinct, the orchids won't be able to reproduce.

In 2023, Katharina contributed to Royal Botanic Gardens Kew's State of the World's Plants and Fungi report. The report found up to 45 per cent of the world's flowering plant species could be under threat of extinction. The family Orchidaceae, the orchids, is one of the most at risk.

'Plants and fungi sustain life on Earth and provide valuable ecosystem services, food, medicine, clothing, and raw materials,' Katharina says. 'But the natural world is threatened by the dual crises of climate change and biodiversity loss.'

'We are working to add threatened orchids from Australia to the International Union for Conservation of Nature Red

List of Threatened Species. Most of them are terrestrial, meaning they grow on the ground, showcasing the rich but threatened diversity of terrestrial orchids found in Australia.'

Some species are so rare that despite a lifetime working on orchids, Mark has never seen the living plants. One of these is *Taeniophyllum cylindrocentrum*, a tiny, tree-dwelling species known from only two specimens. The species made its scientific debut in 1912 in a book about more than 1,500 orchids from New Guinea by a German botanist, who had spent several years living on the island. His specimen of *T. cylindrocentrum* was destroyed when the Berlin Herbarium was bombed during World War II.

To his delight, Mark discovered the second specimen in his laboratory when he decided to identify a mystery specimen on loan from the Queensland Herbarium. It had been collected from Cape York Peninsula in 1978 and resembles a tiny, faded clump of grass floating in ethanol.

There are no photos of this rainforest species growing in the wild. It has no leaves and uses its green roots to both anchor itself to a tree and photosynthesise.

'I would love to see this species in the wild,' Mark says. 'But the location is remote and inaccessible. And there are crocodiles.'

A parrot in the night

Orchids are not the only species to go missing. The Night Parrot (*Pezoporus occidentalis*) is perhaps Australia's most elusive bird, elevated to almost mythical status by its century-long disappearance.

Populations of the Night Parrot were once widespread across the country, but dwindled as pastoralism and predators like feral cats and foxes spread. Between the end of the 19th century and 1990 almost every report of Night Parrots was unreliable. Today, very small populations are known to exist only in south-west Queensland and inland Western Australia.

Unusual among the world's 400 parrot species, the Night Parrot is nocturnal. Its yellow, green and black-flecked colouring is similar to that of its distant cousin, New Zealand's much larger, flightless, nocturnal parrot, the Kākāpō. Although the Night Parrot can fly, it dwells and nests on the ground in spinifex grass, its colours providing camouflage.

In the freezers of the Australian National Wildlife Collection is a precious specimen of muscle tissue from a Night Parrot. The tissue was donated by the Western Australian Museum, from their single Night Parrot specimen that was found deceased by Traditional Owners in the Pilbara in 2022.

After extracting DNA from a tiny piece of this frozen tissue, CSIRO researchers sequenced the full genome of the Night Parrot. The result is a long string of DNA's four-letter code, which has been 'annotated' by mapping the location of known genes onto the string, based on genetic research on other birds.

Dr Leo Joseph, director of the wildlife collection, tells me researchers will use the Night Parrot genome to answer questions about its biology and populations.

'We'd like to find out which genetic changes to vision and navigation have enabled Night Parrots to be nocturnal. This is very unusual in parrots,' Leo says. 'There are also ways to use the genome sequence of a single bird to estimate past population sizes of the species. That will help us understand more about their decline.'

I ask him whether the genome will help save the species, but he says we already know a lot about how to do that.

'Protecting Night Parrots from cats, foxes and habitat loss are the first ways to save the species,' he says.

Now that DNA work is commonplace, bird specimens in collections have matched samples of muscle tissue held in deep freeze. The birds themselves are preserved as study skins. They are stuffed with cotton wool and feel almost as light as their feathers. Study skins are not posed in lifelike positions with glass eyes. Instead, they lie flat, with their wings folded and cotton wool in place of eyes. A thin, wooden rod provides stiffness to the body.

Study skins help bird taxonomists understand the morphology of different species. Their data also record dates and locations, revealing exactly when an individual bird was present in a particular place. This means specimens can reveal migration patterns, which is what Leo and colleagues set out to do by examining specimens of the Black-winged Monarch (*Monarcha frater*).

These insectivorous songbirds live in the rainforests of the island of New Guinea and Far North Queensland (FNQ). One of the four subspecies, *M. frater canescens*, migrates to FNQ from November to March. When the dry season arrives, the birds disappear north to New Guinea, but Leo says no one has any idea exactly where. He wondered whether looking at the 194 specimens of Black-winged Monarchs held in the world's museums and collections would enable him to pinpoint their whereabouts during the non-breeding season.

It didn't. But it did turn up something surprising: the specimens didn't fit well into four subspecies.

'Based on differences in the patterns on their heads, we think there may actually be three separate species of Black-winged Monarchs, not one species divided into four subspecies,' Leo says.

DNA analysis could resolve this. It lets researchers trace intricate evolutionary relationships. Unfortunately, only four of the 194 Black-winged Monarch specimens have frozen tissue stored. But using recently developed DNA sequencing techniques, researchers can now extract DNA from the toepads of bird specimens. Leo's team is pursuing this to test their hypothesis that Black-winged Monarchs are actually three separate species.

In the meantime, they suggested a name for the birds that breed in FNQ – the Pearly Monarch. As for where they go from March to November, Leo paraphrases Oscar Wilde.

'To lose a species is tragic; to not know where it is for half of every year is careless,' he says.

A sponge filled with sharks

Australia's many species of sharks and rays are relatively well known, at least when compared with groups like insects. But familiar species of sharks and rays found in unusual locations can be big news in science.

In the winter of 2022, Dr Will White, a shark expert at the Australian National Fish Collection in Hobart, received an unusual phone call. A devil ray, normally at home in tropical and subtropical waters, had washed ashore at Coles Bay, Tasmania.

Will and colleague Helen O'Neill left work for the four-hour round trip, excited to add a new specimen to the collection. In 2017 Will had published research merging devil rays and manta rays into a single genus, *Mobula*. He identified the devil ray at Coles Bay as a Giant Devil Ray (*Mobula mobular*). It was 2.5 m wide.

Too big to store as a whole specimen, Will and Helen dissected the devil ray and preserved parts of its body for future research, including its gills, eye, skeleton, gut and uterus.

I see one of these specimens for the first time a year or so later when I ask Will and Helen to give a virtual tour of the fish collection to a group of filmmakers. They show us a milky eyeball of the devil ray from Coles Bay. It's the size of a cricket ball and is stored in a jar of ethanol.

'This species is poorly represented in biological collections due to its large size,' Helen says. 'This is an important specimen to have.'

The fish collection functions like a library for research by both the staff here and researchers from around the world. More than 160,000 fish specimens representing 3,700 species are preserved in ethanol.

Smaller fish are stored in individual jars on mobile shelving. Fish lose their colours during the preservation process, which involves fixation in formalin before long-term storage in ethanol. Will shows us an angler fish from the deep sea. A fleshy lure dangles from its head. In the deep sea this would be illuminated to attract prey to its gaping, teeth-filled mouth below. But this terrifying sea monster is tiny enough to sit in the palm of your hand.

Larger fish are stored in ethanol-filled tubs the size of a large bath. Will puts on long green rubber gloves and lifts out a sawfish with a long rostrum, or snout, spiked along its edges with dozens of sharp teeth. In another tub he reveals a

Leopard Whipray (*Himantura leoparda*) with a pattern just like the big cat.

Across the room, Al Graham, the manager of the fish collection, places two glum-looking monkfish on top of their tub. They are specimens he regularly pulls out of the tubs when showing the collection to visitors.

'These fish have been observed on video perched or sitting on rock ledges and rocky outcrops,' he says. 'The water supports their body, so they look more plump when observed underwater. In their preserved state, their body becomes more flattened.'

Some specimens like the devil ray from Coles Bay have entered the fish collection opportunistically. Another example is a Broadnose Sevengill Shark (*Notorynchus cepedianu*) that washed ashore in 2016 on the banks of the Derwent River in Hobart, right outside the fish collection. Staff arriving at work were quick to recognise its value to science and preserved it in the collection.

Many fish in the collection were collected during biodiversity surveys on board CSIRO's research vessel (RV) *Investigator*, such as some sharks found in a very unusual habitat.

In 2017, marine scientists on board RV *Investigator* off the north-west coast of Australia landed a huge sponge on the deck. Poking out was a patterned tail. With gloved arms, they pulled out a small Banded Sand Catshark (*Atelomycterus fasciatus*). It was the first known observation of a shark living inside a sponge, quickly followed by the second, third, fourth ... and 29th, all inside the same sponge.

Across two voyages in north-west Australian marine parks, the team discovered five large sponges at depths of up to 106 m harbouring a total of 57 Banded Sand Catsharks, including males and females of mixed ages and sizes. They kept a few as specimens and the rest were released.

Helen was on board the second voyage and was the lead author of the paper describing the behaviour. She says some of the sponges harbouring the catsharks were up to a metre in diameter, with many folds, crevices and canals where they could hide.

'Some species of bony fish use sponges as microhabitats, but this wasn't known for sharks,' she says. 'We think the purpose might be to avoid predators during the day, but we still have so much learn about shark behaviour.'

A fly in a toilet

Unlike finding a shark in a sponge, you don't have to be a scientist carrying out a biodiversity survey to find a fly in a toilet. Many of the world's fly species lay their eggs in poo so their larvae – aka maggots – can take advantage of this abundant food source.

The Bush Fly (*Musca vetustissima*), the Housefly (*M. domestica*) and the many mosquito species that feed on our blood have made flies hard to love. But for dipterists – scientists who specialise in flies – there are around 25,000 species to lavish their attention on. These two-winged insects are so diverse there are even flies disguised as bees, and they pollinate flowers just as well as bees do.

Keith Bayless is a research scientist at the Australian National Insect Collection. He hunts rare flies to discover biodiversity and understand more about the family tree of flies. Since 2015, he's been on the trail of a very rare species: *Clisa australis*.

This fly was described as a species in the 1960s, after it was found living in a cave near Kempsey, New South Wales (NSW). The flies were associated with bent-wing bats, their larvae feeding on the bats' poo. Over the next couple of decades, scientists spotted *C. australis* around NSW in caves, mine shafts and pit toilets. Then it went missing.

C. australis is the only fly of its kind on the Australian continent. Its closest relatives live in south-east Asia and Lord Howe Island. Visiting the caves where the flies were first spotted was out of the question thanks to steep cliffs and stinging plants blocking the entrance. Then a megafire impacted the reserve in November 2019. Determined to track the fly down, Keith decided pit toilets were his best bet. Humid holes filled with rotting organic material are not too dissimilar to caves filled with bat poo.

'I've checked the walls and ceilings of an embarrassing number of public toilets in national parks looking for *C. australis* adults,' he says.

After years of unsuccessful searching, Keith set up a Malaise trap in remnant rainforest at Barren Grounds, NSW.

Traps to catch flies can be very simple. A yellow plastic bowl filled with water and a drop of detergent will soon be dotted with flies. They are attracted by the colour, and land on the water but are unable to stay afloat as the surface tension of the water is broken by the detergent. A Malaise trap is a little more sophisticated, being a mesh tent large enough for a person to lie under. Flying insects that enter the trap are channelled upwards into a jar of ethanol. Leave it in place for a week and you'll have a grotesque concoction of pickled insects, or as Keith calls it more poetically, a distillation of the forest.

Keith's target for the Malaise trap at Barren Grounds was Notoconops, a genus of parasitic thick-headed flies. (He tells me these top-heavy flies are very distinctive in appearance, but whether notoconopines are true Conopidae is 'noto' so certain.) The target flies never turned up in the trap, but to Keith's great surprise *C. australis* did.

Not hanging out with bats in a cave or loitering in a pit toilet, the missing fly was living beside a stream in a dark, humid gully filled with ferns. A week later and 150 km away, Keith caught a second specimen of *C. australis* in an overgrown, muddy ditch at Lake George near Canberra.

'I don't know whether this species was always widespread but rare, or whether its range and life history are changing due to human-mediated pressures including megafires, habitat destruction and sanitation,' he says. 'Perhaps they were on the move and had to leave their comfortable pits in search of smellier pastures.'

The next time I see Keith, he is just back from a beachside holiday in south-east NSW. The beach had a pit toilet and he couldn't resist checking it. He found males of *C. australis* on the walls and they appeared to be doing a mating dance. What's more, they looked a little different from the specimens he caught in the Malaise traps.

'They might even be a new species,' he says.

The realisation that I've also been discussing parasitic wasps with Ian, at the top of this chapter, sets the dipterists like Keith (who study flies) against the hymenopterists (who study wasps, bees and ants) in a battle for whose group is more impressive.

Keith tells me many groups of flies parasitise insects and spiders. Some fly species even target larger creatures like barnacles and rodents. Bot flies target humans and other mammals, though the result is less spectacular than in *Alien*. The beautiful bee flies that mimic the striped patterns of bees are parasitoids. And in a cruel twist, some bee flies parasitise the larvae of bees. Other flies target the larvae of wasps. We've come full circle in the game of flies versus wasps.

> *Scientists like Girault, dreamer of a wasp on Jupiter, are extreme exceptions in the history of taxonomy, the science of describing and naming species, something I'm keen to learn more about.*

3
Marvellous names

A platypus is actually a weevil, This is the name of a fly, Australia's River Red Gum is named after a monastery in Tuscany, cerulean blue frogs are green in real life and Stan Lee is a robber fly who leads a cast of robber fly superheroes from the Marvel Universe.

What's in a name?

On an autumn day, I take my Dad along to the Australian National Wildlife Collection to photograph a bird specimen from the Solomon Islands that has been lodged there. As we arrive, a large family of White-Winged Choughs (*Corcorax melanorhamphos*) forages in the leaf litter outside the building. It would be a short field trip if someone at the collection decided to study them.

Inside the building, the beautifully prepared study skin of the bird we're here to photograph lies on a huge table that takes up most of the room. Today, the rest of the table is devoted to curating eggs. There are tiny carboard boxes, paper labels, rolls of cotton wool, scissors, forceps, glue, paper clips and many trays of eggs. They range from delicate, blue egret eggs to large, cream-coloured pelican eggs. I notice elegant pencil handwriting on one of the pelican eggs. It surrounds the small, neat hole through which its contents were removed and reads: Nov 1, 1909; Foster Is.; J.DM^{C}L; 651; 1/2. All of the eggs in the wildlife collection are from historical collections, donated to CSIRO after egg collecting was made illegal in the 1960s.

I start to arrange the bird study skin against a white background. It's a rail, which are water-loving birds named for the sound of their calls, from the Latin word *rādere*, meaning to scrape. The label tied to its legs records that it was found dead on a road on the island of Malaita, a province of the Solomon Islands. The label also notes that there are separately stored specimens prepared from the same bird: 'Prep(s): skin; right wing; eyes + body in spirit'. We also need to photograph the right wing, which has been spread and preserved with its feathers fanned out to reveal their structure, colours and patterns. An extra label reads HOLOTYPE. This specimen is being used to describe and name a species that will be new to science.

Scientific naming is a process that began formally in Europe nearly 300 years ago led by Swedish scientist Carl Linnaeus, who championed the idea of giving every species on Earth a two-part, Latinised name. He named our own species *Homo sapiens* – wise human. The first part of a scientific name indicates its genus, which it may share with similar, closely related species. The second part indicates its specific name, which is always written together with the genus or its abbreviation, as in *H. sapiens* or the infamous *T. rex*.

Today, naming a new species requires three things: a taxonomist or a team of taxonomists, a holotype and a scientific paper. The holotype is the specimen chosen to anchor the species name. In collections they are often closely guarded, kept in fire-safe rooms and labelled HOLOTYPE in bold letters.

Scientific naming of species doesn't erase other names for plants and animals used by Indigenous Peoples or in the vernacular, nor do they necessarily overlap. For example, a single plant species having one scientific name may have multiple names based on its different uses or in different languages. Scientific naming is a separate system that is international and based on the taxonomic relationships between species. These days the scientific papers that describe and illustrate new species often include a phylogenetic tree, which is a branching diagram that shows how different species are related to each other based on their DNA.

Around 1,000 new species are scientifically named in Australia every year. The majority are insects, but new fishes, birds, plants, fungi, algae, marine invertebrates and even lizards and mammals make the list. Some new species are entirely new discoveries, such as never-before-seen creatures collected from the deep ocean. Others are simply being recognised scientifically for the first time. A new species of mosquito from a forest in Victoria might have waited in the insect collection for 50 years for a taxonomist to have the time, skills and funding to study it. A new species of bird from Cape York Peninsula might finally have been revealed, through study of its plumage patterning and DNA analysis, to be different enough from its cousins in New Guinea to be named as a separate species.

Choosing a name

New species are often named for their features, especially their appearance. *Ogyris caelestia* is a new butterfly species from Queensland that was named in 2023. It belongs to the genus *Ogyris* along with 17 closely related butterfly species. Its species epithet, *caelestia*, was chosen by the taxonomists who wrote the 41-page paper describing it in minute detail all the way from its wings to its genitalia (see Chapter 8). *Caelestia* is

Latin for 'of the sky, or celestial', calling to mind both the bright blue upper wings of this species and the way it flies above the tree canopy. Taxonomists may also give a new species a formal common name, which stands in place of the species name in everyday use. In this case they chose Sapphire Azure.

In the wild, this butterfly depends on a new species of ant that was named in 2021: *Anonychomyrma inclinata*. The two species live together in an arrangement with a long name: an obligate myrmecophilous relationship. The ants care for the caterpillars of the Sapphire Azure butterfly and an unknown number of other butterfly species that live in trees and feed on mistletoe. They protect the caterpillars from predators and carry them to mistletoe leaves to feed. In return, they receive a sugary reward secreted by the caterpillars. This ant species also attends larvae of the Bulloak Jewel Butterfly (*Hypochrysops piceatus*), a threatened species that lives in southern inland Queensland.

The ant's species epithet, *inclinata*, is from the Latin *inclina*, meaning inclined. According to the authors of the paper describing it, the name refers to 'the slope of the propodeal dorsum of the worker' compared with that of similar ant species. It's an important feature used to tell these ants apart from similar species, but somewhat cryptic unless you're in the know.

Naming a species for obscure features is quite common. Take *Boea resupinata*, for example. It's a rainforest herb that grows in a few areas on Cape York Peninsula. It has purple flowers and looks a bit like an African violet. That's because they're in the same family, Gesneriaceae. *Boea* species are often called resurrection plants because they can dry out and rehydrate when it next rains. *B. resupinata* was named as a new species in 2021 by Frank Zich at the Australian Tropical Herbarium in Cairns, together with one of his colleagues. But the first specimens were collected way back in the 1970s.

'It wasn't recognised as being a new species until I was curating specimens in the herbarium and recognised that is was distinct from its Australian relatives,' Frank says. 'The name we chose is from the Latin *resupinus*, meaning bent backwards, referring to the twisting of flowers through about 180° as they open.'

The rule of thumb is not to name a species after yourself, but you can name one after someone else. The entomologist Dr Bryan Lessard named the soldier fly *Apisomyia bathae* after the journalist Chris Bath, to thank her for featuring him regularly on her ABC Radio show to tell the world about flies. The fly takes a Latin form of her surname. Parasitologist Dr Daniel Huston named the marine trematode *Enenterum petrae* after his baby daughter, Petra. It's a worm-like parasite that lives inside a species of marine fish, *Kyphosus vaigiensis*, the Brassy Drummer. The parasite takes the possessive feminine Latin form of the daughter's first name, meaning 'of Petra'.

Species are sometimes renamed when there's been some kind of taxonomic confusion. The name *Platypus*, which means flat foot, was given to a genus of Australian weevils by a German entomologist in 1793. *Platypus* weevils and related species bore into wood, creating tunnels in which they farm fungi for their larvae to eat. In 1799, a naturalist at the British Museum named Australia's strange-looking aquatic mammal *Platypus anatinus*, which means flat-footed duck. A few years later the mistake of using a genus name twice was realised.

Being first, the weevil won. The mammal was renamed but the common name Platypus stuck.

In 1991 an entomologist at the Australian Museum in Sydney named a genus of flies *This*, not just to be funny. (Have you seen *This*? I can't find *That*.) *This* is derived from an Ancient Greek word meaning seashore. The larvae of *This* flies feed on kelp that has detached and washed ashore at the end of its life. They play an important ecological role by breaking down rotting seaweed and recycling nutrients. So far there is only one species in this genus, *This canus*. This fly lives along the southern half of Australia's coastline.

Australia's River Red Gum (*Eucalyptus camaldulensis*) was named in 1832. The suffix -ensis at the end of a species name indicates it's named after a place. But this iconic tree that grows along inland waterways throughout Australia took its name from a place far from where it naturally occurs. It was planted in a garden near the Camaldoli monastery in Italy in 1818. In 1832 it was named as a new species by a landscaper in a booklet that listed all the plants in the monastery garden, *Catalogus Plantarum Horti Camaldulensis*.

The Green Tree Frog illustrates how even naming a species after its looks can go wrong. By the time specimens of this Australian species were in the hands of the scientist who named it in 1790, the preservative they were stored in had transformed their bright green skin to a lovely cerulean blue. The scientific name of the Green Tree Frog is *Litoria caerulea*.

Distant cousins in the sea

Sharks, skates, rays, sawfishes and chimaeras are literally in a class of their own. It's called Chondrichthyes, meaning cartilaginous fishes – creatures whose skeleton is made of a tissue called cartilage, not bone. Taxonomically speaking, a shark is as closely related to an elephant as to bony fishes such as tuna. Put another way, a tuna and a human are more closely related to each other than either is to a shark.

In 1999 a team of scientists including Dr Peter Last of the Australian National Fish Collection were working in the Arafura Sea north-west of Darwin when they netted something much larger than their target organisms, which were tapeworms that inhabit the intestines of Australian fishes. They caught a giant stingray, 1.6 m across, with a long, whip-like tail, a finely blotched pattern and yellowish-grey colouring. It was a species of whipray that no one recognised. They took photos and a tissue sample and returned the whipray to the sea.

A decade would pass before DNA analysis of the tissue revealed the whipray was an unknown species. But it couldn't be named. It's necessary to have a specimen when scientifically naming a species. The peril of using a photograph was illustrated in 2022 when a group of scientists wrote a paper extending the known range of a species of Goblin Shark using as evidence a photo taken by citizen scientists on a beach in Greece. Concerns were raised by shark experts around the world, including Dr Will White of the fish collection, who pointed out the unexpected features and location of the Goblin Shark and noted its strong resemblance to a children's plastic toy. The paper was retracted. It was a cautionary tale for scientists and a metaphor for a marine environment cluttered with plastic pollution.

The mystery whipray spotted in the Arafura Sea began to reveal itself during surveys in Kakadu. Although only tissue samples and photographs could be taken, this allowed DNA analyses to confirm the whiprays were all a match to each other. Eventually, ray biologist Dr Peter Kyne was able to collect a single specimen of the new whipray from the Cambridge Gulf of the Kimberley region and lodge it in the Western Australian Museum's fish collection.

Around the same time, Will spotted the mystery whipray in photos of by-catch from the Gulf of Papua prawn trawl fishery. This placed the species right across the north of Australia to Papua New Guinea.

Peter Last, Will White and Peter Kyne wrote a paper describing the species using the specimen in the Western Australian Museum and included photos of 15 other specimens. They named it *Urogymnus acanthobothrium*. Its species epithet borrows the name of a genus of tapeworms, *Acanthobothrium*, first found inside a species of whiptail stingray from the Gulf of California. Like a snake swallowing its tail, the mystery whipray first noticed while studying tapeworms is named after a tapeworm that lives in a whipray.

How had one of the world's largest species of stingrays been overlooked by science for so long? It's not just rays that are hiding in plain sight. There are newly named species of sharks swimming in Australian waters.

In the waters west of Ningaloo Reef in Western Australia lies Ningaloo Marine Park. In November 2022, RV *Investigator* sailed to the area to survey seabed habitats and biodiversity. Finding an unnamed species of shark among the creatures observed wasn't much of a surprise. DNA studies had already revealed the presence of an extra species related to the Port Jackson shark in the order Heterodontiformes, a name that suggests these sharks share some kind of noteworthy dental feature. Helen O'Neill, a fish biologist at the fish collection, tells me species in this order have rows of familiar-looking shark teeth at the front of their jaws. Further back they have rows of molars.

'This dentition is unique among sharks,' Helen says. 'They also have a unique body shape and horn-like structures that are formed by crests just above their eyes. These sharks sit on the sea floor and feed on creatures like molluscs and crustaceans. Their mouths are small but they have strong jaws powerful enough to crush cowrie shells.'

By the time Helen was on the voyage to Ningaloo Marine Park, genetic studies of specimens of Heterodontiformes held in different collections had revealed that one of the nine species in the order, *Heterodontus zebra*, was actually two species, bringing the tally to 10.

'By this time we already had enough information including specimens, genetics and imagery to write a paper describing and naming the new shark,' Helen says. 'But in Ningaloo Marine Park we collected a specimen that became the holotype, meaning the name-bearing specimen that provides the description or blueprint for the species. We named it *Heterodontus marshallae*, the Painted Hornshark' (Plate 5).

The name *H. marshallae* honours artist Dr Lindsay Marshall, who painted every living species of shark and ray for the Chondrichthyan Tree of Life project and the CSIRO *Rays of the World* book – a total of more than 1,200 species.

There turned out to be around a dozen specimens of *H. marshallae* in museums and fish collections around Australia. The specimen collected during the voyage was a male.

'We prefer to use males for shark holotypes because they have claspers, which are external reproductive organs that can vary between species and help us tell them apart,' Helen says.

As for its whereabouts in the wild, *H. marshallae* lives only off north-western Australia at depths of around 125–229 m. *Heterodontus zebra*, the species it was previously lumped with, lives in Indonesia and northwards to Japan in shallower waters, ranging from the surface down to 143 m. With such different distributions and depths, why were the two species confused for so long? Helen says their similar colours and markings were to blame. Genetically, the Painted Hornshark is more closely related to the Port Jackson Shark.

Such galahs

Birds are close to winning the naming race. There are roughly 800 named species in Australia and that's likely very close to the final tally. Among the most recognisable are galahs, the pink and grey cockatoos that occur across most of the continent. But until 2016 galahs had their names in a muddle – or rather ornithologists did.

Dr Leo Joseph, Director of the Wildlife Collection, tells me galahs were given a scientific name in the early 1800s and today are known as *Eolophus roseicapilla*.

'The holotype of the species, which anchors the scientific name, is held in Paris in the Muséum national d'Histoire naturelle', he says. 'It was collected in 1801 by biologists on France's Baudin expedition.'

The Baudin expedition was signed off by Napoléon Bonaparte to map the coast of the country they called Nouvelle Hollande and explore its natural history. It departed France in October 1800 and at the end of a three-and-a-half-year journey arrived back home with 200,000 specimens, but only a fraction of the scientists. The rest had left the expedition due to illness, or died at sea. Despite the loss of people on a journey typical of its time, more than 3,000 species were scientifically named as a result of the Baudin expedition.

Eventually, ornithologists split galahs into three different subspecies, adding a subspecies epithet: *E. roseicapilla roseicapilla* for galahs in the east of Australia, *E. roseicapilla assimilis* in the west and *E. roseicapilla kuhli* in the north.

The holotype in Paris was assumed to be an eastern galah and thus owned the name *E. roseicapilla roseicapilla*. But was this assumption correct? Was the holotype in Paris really an eastern galah?

'Galahs have dramatically expanded their range since colonisation,' Leo says. 'Eastern galahs didn't overlap with the route of the Baudin expedition. They do now, but they didn't back then. My predecessor Richard Schodde raised the alarm here and we followed it up.'

The other clue was in the species' name.

'Eastern galahs have pale, almost white heads,' Leo says. 'But roseicapilla means rosy hair. Western galahs have rose-coloured heads.'

I ask Leo why he didn't just take out the holotype and look at whether its head feathers were pink or white. He reminds me that it's in Paris. It's also more than 200 years old and isn't in great shape. However, it does appear to have a rosy-pink head. How do we get to the bottom of this question? Let's enter the world of ancient DNA.

Bird skins, or study skins, in collections usually have matching tissue samples stored at −80°C. Frozen tissue is a ready source of DNA. But skins from centuries ago can also be a source of DNA. So Leo took a tiny snip of skin from the toepad of the holotype in Paris and his team used techniques that work for sequencing degraded DNA. It was a match for western galahs.

I ask Leo what happened to the names.

'It meant western galahs own the name *E. roseicapilla roseicapilla*,' he says. 'Richard Schodde gave eastern galahs a new name, *E. roseicapilla albiceps*, which refers to their white crowns. The holotype for that name is here in our collection.'

The muddle of galahs is one of many stories Leo has about bird names. But because birds are so well known compared with groups like insects, where only around a quarter of the species have names, bird taxonomists tend to focus less on describing new species and more on asking deeper questions in evolutionary biology. Where did songbirds first arise before they spread around the world? Does coevolution between cuckoos and their hosts lead to new cuckoo species arising? How many species of long-extinct elephant birds existed on the island of Madagascar?

Stan Lee and his insect assassins

Insects dominate the tally of newly scientifically named species in Australia every year. With so much diversity in morphology and the huge number of insect species yet to be named – more than 200,000 – it's almost inevitable that taxonomists will spot doppelgangers among them. Recent finds in Australia include *Leptanilla voldemort* – Harry Potter villain Voldemort masquerading as an ant; *Opaluma rupaul* – a soldier fly dressed in an iridescent gown belonging to RuPaul, host of *RuPaul's Drag Race*; and *Paramonovius nightking* – a bee fly named after the Night King from *Game of Thrones*.

3 – Marvellous names

When Isabella Robinson was an undergraduate entomology student, she took on a project at the Australian National Insect Collection to scientifically name some new species of robber flies. While looking down a microscope at the magnified details of their eyes and legs, thoraxes and antennae, she noticed a curious-looking robber fly.

Robber flies are the assassins of the insect world. Their hunting style is to lie in wait and ambush other insects as they fly past. Flies in this family, Asilidae, are long and thin with bristly bodies. In one new species in the genus *Daptolestes*, those bristles and the robber fly's large eyes gave the creature a strong resemblance to the bristly moustache and sunglasses of Stan Lee, co-creator of the Marvel Universe. Isabella named it *Daptolestes leei* in his honour. Getting creative, she named three other flies in the same genus after Marvel characters: Loki, *Daptolestes illusiolautus*, meaning elegant deception; Black Widow, *Daptolestes feminategus*, meaning woman in leather; and Thor, *Daptolestes bronteflavus*, meaning blond thunder.

The final robber fly was a dead ringer for Deadpool, with markings on its back that have an uncanny resemblance to Deadpool's mask (Plate 6). For this species, Isabella also needed to name a new genus, the level up from species. She came up with *Humorolethalis sergius*. The genus name sounds like lethal humour, derived from the Latin words *humorosus*, meaning wet or moist, and *lethalis* meaning dead – a fitting name for a comedic assassin.

Un-naming

A chapter on naming species would not be complete without a mention of its counterpoint: synonymising. Enter the bizarro world of un-naming.

At times the efforts to describe and name the 10 million or so species on Earth (not counting microbes) lead to a single species being named as two separate species. This often comes to light years later when DNA evidence is added to the picture for the first time.

The process to synonymise two species into one is much the same as naming a new species: a taxonomist or a team of taxonomists examines specimens and writes a paper.

To find out more, I visit Dr Cecile Gueidan, a lichenologist at the Australian National Herbarium. She tells me lichens can look dramatically different, even though they're actually the same species, sort of like identical twins with different hairstyles.

'Until very recently, lichens were named as new species based on their morphology, meaning how they look, as well as the unique chemicals they produce,' she says. 'But lichens can look very different depending on the substrate and environmental conditions in which they grow. This means two named species can sometimes end up being a single species.'

Cecile is using DNA techniques to reveal how Australian lichen species are related to each other. As a result, in 2022 she collapsed a total of seven lichen species into just three species. Two of the species that were synonymised were *Trapelia pruinosa* and *Trapelia rosettiformis*, both named in 2020. Cecile says the specimens actually look quite similar, until you get very close.

'They seemed to differ by the texture of their upper surface, the size of their spores and the presence of calcium oxalate crystals,' Cecile says. 'But our DNA data showed they are likely to be a single species.'

I ask Cecile how she decided what to name it. But that wasn't her decision to make.

'The name *Trapelia pruinosa* was published first, so the rule is we keep that name,' she says.

Piecing together the family tree of lichens is just a first step in Cecile's work. Her goal is to hunt for new resources in nature, something many researchers at collections are interested in.

4
Cocktails in nature

In a dark hole, a little egg lay in the body of a paralysed spider. One morning the warm Sun came up and – pop! – out of the egg came a tiny and very hungry larval wasp. It didn't need to look for some food. The spider was its larder, kept fresh by a cocktail of venoms. These molecules, alongside new products from algae and lichens, could soon join the plethora of pharmaceuticals and food products sourced from nature.

Cocktails of spider wasps

I arrive home from work one day to find a large wasp attempting to drag a huntsman spider up my front door. I assume it's heading to a gap between the bricks above the door, but the spider is heavy and the door is high. The wasp makes three failed attempts before dropping the paralysed body of the spider on my doorstep and flying towards me. I jump out of the way. Spider wasps (Plate 7) rate four out of four on the Schmidt pain index, a scale invented by an entomologist who tested the stings of Hymenoptera – ants, bees and wasps – on himself. I've no desire to check his results.

Spider wasps range from ~5 mm to 60 mm, each species specialising in hunting different species of spiders. The female wasp prepares a burrow, hunts her preferred spider, paralyses it with her sting, drags it into her burrow and lays an egg inside its body. Her venom not only paralyses the spider, it keeps it alive and fresh while the larval wasp devours it from the inside, saving the central nervous system for dessert. It then pupates and emerges as an adult wasp.

I upload a photo of the wasp that was on my front door into the iNaturalist app, identify it as a Zebra Spider Wasp and email Dr Juanita Rodriguez, a hymenopterist at the Australian National Insect Collection. She responds with wry humour that this is why I should carry an entomologist's net everywhere I go (Plate 8).

Juanita works on Pompilidae, the spider wasp family. They're the larger cousins of the parasitic wasps we met in Chapter 2, the 'tea leaves' in Dr Ian Naumann's office. Researchers like Juanita are interested in their venoms as a source of new pharmaceuticals. I visit her office, where she has drawers of spider wasp specimens spread out.

'Spider wasps are very interesting because they have not been studied very well in Australia. There are a lot of new species that we need to describe,' Juanita says. 'It's not only the diversity of species that we're looking at. We're also looking at the diversity of venoms they produce.'

'Spider wasp venom is a cocktail of hundreds of molecules. Some of them have been found to have an effect in sodium channels that are involved in neuronal function. Because of this effect, we think they may be useful to treat conditions like Alzheimer's, epilepsy and Parkinsons.'

It's amazing to think a venom molecule from a cocktail that nature invented around 40 million years ago could find its way into a pill.

I pop down the hallway to ask Ian what he knows about spider wasps and he tells me yet another version of the life cycle story: wasps that parasitise egg sacs.

'Spider egg sacs contain many spider eggs, maybe five to 50,' he says. 'The female wasp lays several eggs inside the egg sac and the wasp larvae hatch and feed on multiple spider eggs. The wasp larvae are like little predators inside the egg sac. I recall reading once about a wasp that attacks the egg sacs of redback spiders and has this kind of life cycle. I liked that story because I am not so keen on redbacks.'

The language of lichens

Next, I continue my visit to the laboratory of Dr Cecile Gueidan at the Australian National Herbarium to learn more about her research on lichen products. Her office is filled with books, journals, a microscope, boxes of specimens and a container of glass vials filled with coloured powders.

A lichen is a symbiosis between two very different lifeforms: a fungus and an alga. Worldwide, there are more than 20,000 different fungi species that form lichens by pairing with around 300 different algae species. Scientifically, they are named after the fungus in the partnership. How do fungi and algae cells find each other in the environment? Nobody knows, and lichens are incredibly difficult to cultivate in the laboratory. But if scientists can get on top of that, lichens might one day be good candidates for growing on Mars.

Cecile tells me that although the word 'symbiosis' suggests lichens are an equal partnership, the fungi may actually be farming the algae. Fungi normally feed by absorbing nutrients from the environment, like organic matter in soil or rotten fruits. By farming algae, lichenised fungi can feed on the sugars algae produce by photosynthesising. They have a ready source of food, as long as the Sun is shining.

Before scientists like Cecile began to use DNA to tell lichens apart, the biochemistry of lichens was an important feature for identifying them.

'You can use simple chemicals like bleach and potash to tell some lichens apart. For example, a drop of bleach will turn some lichens pink,' Cecile says.

This is something I am keen to see, so Cecile opens a specimen box and removes a twig covered in a pale green lichen. Working under a microscope, she scrapes off a tiny piece of the lichen's surface with a scalpel and adds a tiny drop of bleach. It turns bright pink for a few seconds and then fades.

Cecile has collected some products containing lichens, which she talks about during tours of the herbarium. One is a tube of toothpaste that claims to whiten teeth, thanks to the lichen extract it contains. Another is a tin of solid perfume extracted from oak moss. It has sweet and earthy notes, like eating a crème brulée in a forest. Oak moss is not moss, it's a Northern hemisphere lichen that covers trees such as oaks in pale green clumps.

Lichens grow on the surfaces of various substrates, including bark, rocks and concrete. Broadly speaking, they can be divided into three different groups: crustose lichens, which form tightly attached crusts; foliose lichens, which form looser, slightly raised structures; and fruticose lichens like oak moss, which grow in branching, spindly formations. Oak moss has been harvested for its essential oil for centuries. Before the invention of synthetic dyes, lichens were also used to dye fabrics, such as Harris Tweed.

I'm curious about the smell of oak moss. Why does this lichen smell so wonderful to humans? Is oak moss trying to attract beneficial insects through scent? This is a well known strategy in the world of flowering plants, but it turns out a flower can do more than look pretty and smell sweet to attract pollinators. There's also mind control.

Caffeine is produced by some plants to prevent herbivores eating their leaves and seeds, possibly by causing unpleasant effects that feel like drinking way too many cups of coffee. But perhaps plants also produce caffeine to manipulate their insect pollinators. A study published in the journal *Science* in 2013 found that *Coffea* and some *Citrus* species produce caffeine in their nectar at levels below the ability of honey bees to detect bitterness. Laboratory experiments showed caffeine at these doses helps honey bees remember floral scents and associate them with a reward. This suggests some flowers give their pollinators more than just sugar: they spike their drinks to keep the pollinators coming back for more. It's a chicken and egg question to ask which came first in the evolution of caffeine production by plants: keeping herbivores at bay or using mind control over pollinators. Perhaps the answer is some other effect of caffeine in nature. But for the purpose of people using the plant, the point is simply that caffeine is a neuroactive compound that keeps us alert.

Around three-quarters of pharmaceuticals are sourced from or inspired by nature. Famous early examples include quinine extracted from the bark of *Cinchona* trees to treat malaria; salicin in willow trees, which led to the development of aspirin; and penicillin, the first antibiotic, which was extracted from a mould. Could lichens be next?

Cecile refers to the unique biochemicals that lichens produce as secondary metabolites. Primary metabolites are the essential biochemicals a lichen produces to grow and reproduce. Secondary metabolites are bonus extras that help lichens survive in nature.

'Lichens are slow-growing,' Cecile explains. 'They need to compete with other lichens, fight off microbes and avoid being eaten by animals. Secondary metabolites help them do these things.'

Cecile is interested in secondary metabolites for the effects they might have on the human body. She is referring to the search for new pharmaceuticals.

I peek into the box of vials on Cecile's desk. They have a vintage look, with handwritten labels browned by time. Each holds a small amount of coloured powder, some white, some orange and others a vivid yellow. They are secondary metabolites extracted from lichens by Jack Elix, a retired chemist at the Australian National University, who donated his lichen and lichen compound collection to the Australian National Herbarium.

'We're using mass spectrometry to help understand the nature of these lichen compounds,' Cecile says. 'We're also testing the bioactive properties of extracts of some lichens, such as whether they contain antibiotics.'

It's a story playing out in many laboratories around the world.

Diving into algae

It's December 2023 and I'm having an I-can't-believe-this-is-my-job moment that lasts all week. I'm at a conference that feels like a mini break with 70 or 80 scientists who work on environmental genetics. We're staying in bushland with views across Sydney Harbour. There are cicadas singing in the trees and Long-nosed Bandicoots hopping near our cabins at night. This is the old quarantine station that operated for 150 years, isolating shiploads of immigrants before they mixed with the established population. Post-COVID it sounds bleak, but many of the old photos show holiday scenes. I wonder if my ancestors

4 – *Cocktails in nature*

stopped here, exchanging cleaning chimneys and sculleries in London for this warm, wild place.

Before and after the workday, we go on walks, spot wildlife and play croquet on the edge of a cliff, which is almost as impossible as if the mallets were flamingos and the balls were hedgehogs. In the evenings there are ghost tours of the old buildings for tourists, but my colleagues are too sensible for such things and my attempts to joke about hauntings fall flat. (I manage to snap a photo of Australia's most haunted toilet and I check it later for apparitions. No result.) Each morning the staff from the Australian National Algae Culture Collection go snorkelling in the bay. They are trained divers, who sometimes collect seaweed as part of their jobs. We'll learn more about this in Chapter 10.

One night at dinner during the conference, I'm sitting near the algae researchers when the dessert course arrives, comprising two options that are served alternately. Cheese plates land either side of me, meaning I get the sweet dessert served with ice cream. The algae researchers tell me there's algae on my plate and probably on my face as well. Products from algae are common in foods and toiletries; for example, there are alginates from seaweed in ice cream and algal antioxidants in face creams.

Algae are a mixed group of organisms including seaweeds (macroalgae) and microorganisms such as cyanobacteria, dinoflagellates, diatoms and more (microalgae).

Dr Cintia Iha is a research scientist at the algae collection whose expertise is seaweed taxonomy. She is using AI to search for bioproducts in algae.

'Algae are very diverse,' Cintia says. 'They make many high-quality natural products with different functions.

Examples include oils that are useful for industry, pigments that can be used in paints and food, and sunscreens made by seaweeds to block UV light.'

I ask Cintia how AI is involved in the search for products in algae.

'We are using huge amounts of data from genomics, proteomics and metabolomics,' Cintia says, referring to information about DNA sequences and proteins, oils and secondary metabolites produced by algae.

'The AI will combine the data, sort it and use it to predict the presence of interesting bioproducts in particular species based on their relationships.'

A living collection

Walking around the algae collection in Hobart, the colours of the microalgae cultures are its most striking feature, notwithstanding the view at the end of the building where the laboratory benches look out across the Derwent River at the point in the estuary where the annual Sydney to Hobart yacht race finishes.

Unlike the wildlife, insect, plant and fish collections, the algae collection is a living collection. Around 100 cultures of seaweeds float in flasks and test tubes; tiny compared with the tangles of seaweed that wash up on beaches. More than 1,000 strains of microalgae grow in flasks of culture media or are streaked across plates of solid agar. They are stored in controlled-temperature rooms that range from 15°C to 25°C. Strains from the Southern Ocean live in a small cabinet at 10°C and those from Antarctica share the 4°C cold room with refrigerated laboratory supplies.

Microalgae come in vibrant shades. Motile microalgae, meaning strains that can move, swim in their culture media, turning the liquid in the flasks and bottles a uniform colour. Others settle in a colourful layer at the bottom of their flasks. Most microalgae photosynthesise. They are coloured by the pigments they use to harvest light, including green chlorophylls and yellow, orange and red carotenoids.

Bellerochea forms a golden-brown layer at the bottom of its culture flasks. It's a diatom, a kind of algae with a cell wall made of silica, like a glass shell. The flasks of *Dunaliella tertiolecta* are a bright lime green. It's one of six strains the algae collection supplies to oyster hatcheries to grow as feedstock for larval oysters. In the wild, larval oysters eat phytoplankton, which are microalgae, during the free-swimming stage of their life cycle, before they settle down and mature into adults.

Ros Watson has worked as a technician at the Australian National Algae Culture Collection for more than 10 years and is my go-to person for fun facts about microalgae. One of my favourites is the colour-changing birdbath algae, *Haematococcus*.

'When the cells are happy, they are green and have two flagella, which are like little tails, for swimming around,' Ros says. 'But when they are stressed because of increased light or salinity, they form a cyst cell and turn pinkish-red due to a carotenoid pigment called astaxanthin.'

'Astaxanthin is a useful antioxidant. It's also added to fish feed to turn the flesh of farmed salmon pink. *Haematococcus* can cause birdbaths to turn pink, hence their nickname "birdbath algae".'

Under the microscope, the biodiversity of the algal world becomes even clearer. Different strains resemble spheres, filaments, snowflakes and strands of beads.

Dr Anusuya Willis, or Sui for short, is a phycologist – a research scientist who studies algae – and the director of the algae collection. Part of her role is sourcing new strains of seaweeds and microalgae for the collection. Being in Hobart, the collection is close to remnant populations of Giant Kelp, undescribed species of coralline algae, which are red seaweeds, and myriad microalgae strains living in the coastal and estuarine waters.

In the summer of 2019, Sui collected a sample of sea water from Pipe Clay Lagoon near Hobart and took it back to the laboratory. Under a microscope she spotted some *Haslea* cells among grains of sand. *Haslea* is a diatom with thin, tapered cells that are about as long as a human hair is thick. It has a unique feature: its cells are blue.

'The name of the pigment these cells produce is "marennine",' Sui says. 'It's an antioxidant named for Marennes-Oleron, a region in France that produces oysters, which turn green when *Haslea* is present in high numbers. They are a delicacy known as *huitres vertes* in French cuisine.'

With the help of a microscope and a micropipette, Sui isolated individual cells from the sea water sample and began to culture them. They reproduce asexually, meaning a single cell can give rise to a new culture strain in the collection.

'We provide algae for research around Australia and the world,' Sui says. 'Our collection allows researchers to address many different topics and compare different types of algae, so it is important that we keep expanding our collection with different species from across Australia's unique biodiversity and environments.'

Part of Ros's job is maintaining continuous cultures of all the microalgae strains.

'We need to transfer each strain to new culture media every six weeks on average, because the cells run out of nutrients,' Ros says. 'We handle one culture at a time to prevent cross-contamination.'

The oldest strain in the collection is CS-29, the diatom *Phaeodactylum*. Under the microscope, it looks like long rods with slightly fattened centres. It was isolated in 1910 and acquired by CSIRO in 1961, but this strain was nearly lost during World War II.

'There's a paper from the 1940s that describes how the laboratory in London where this strain was kept was bombed during the war,' Sui tells me. 'They saved a drop of culture from a broken flask.'

'I studied *Phaeodactylum* for my PhD, investigating cell adhesion for biofouling. This happens when diatoms stick to boats and secrete mucilage, forming a layer ~2 mm thick that can decrease the fuel efficiency of vessels. Their fluorescence can be picked up by satellites and is a concern for security, such as for submarines.'

The prettiest strain in the collection is *Noctiluca scintillans*, which is also known as sea sparkle. It's a bioluminescent algae that flashes when disturbed, possibly to deter predators. When large blooms form, sea sparkle lights up the sea surface with a spectacular electric-blue glow.

'Sea sparkle eats other algae and doesn't photosynthesise,' Sui says. 'The cells are big enough to see with the naked eye. They look like little bubbles.'

Ros leads the way into a back room where microalgae strains are being tested for CO_2 uptake for use in carbon storage. Inside a temperature gradient table, tubes of pink and green algae bubble under different concentrations of CO_2, which is pumped through their culture media (Plate 9). Microalgae have a high rate of CO_2 capture and may be used in bioreactors to capture CO_2 from the atmosphere.

Fossil eyes and plastic munching

While organisms like algae make products that can be isolated and put to use, nature can also be a source of inspiration for new products.

One example is the eyes of caddisfly fossils that date back to 11–16 million years ago, in the middle of the Miocene period. Caddisflies are small, moth-like insects with hairy wings that spend most of their lives as aquatic larvae. The fossils were found at McGraths Flat in central New South Wales (NSW), which is a Konservat-Lagerstätte, meaning a fossil site that has preserved extremely fine details.

Dr Michael Frese is a visiting scientist at the Australian National Insect Collection. He worked with Dr Alice Wells, also a visiting scientist at the insect collection, and Dr Matthew McCurry from the Australian Museum to study the caddisflies at McGraths Flat.

'The fossiliferous rocks of McGraths Flat were deposited in an oxbow lake near a river,' Michael says. 'In that lake, the water was still, and fine, iron-rich sediments allowed detailed fossils to form.'

Michael talks me through a set of images of the caddisfly fossils he took using a scanning electron microscope. In larvae we can see details of their silk glands, mouthparts, gastrointestinal tract and claws and, in adults, details of genitalia.

'We even found pollen inside the gut of one larval caddisfly,' he says. 'But most remarkable are the two fossils of adults that show the corneal nanocoating. This is a very small structure on the surface of the caddisflies' compound eyes.'

Michael thinks they are among the best-preserved fossilised eyes on the planet. Their coating could inspire new products, such as antimicrobial paints or sunglasses that repel water. For the caddisflies, the nanocoating suggests the eyes were anti-reflective, perhaps as an adaptation to low light.

'The surface was super hydrophobic, meaning it strongly repels water, both chemically because it was a mixture of protein and wax, and also structurally,' Michael says. 'Its pattern might inspire new coatings for windscreens or sunglasses. The surface is also anti-bacterial because it stresses the outer membrane of certain bacteria. The information could be used to engineer paint that would make hospital walls resistant to bacterial growth.'

As well as inspiring new products, tiny things in nature can be put to work. If you have a compost bin, you're already using microorganisms to recycle food and garden waste. Just as they do in nature, microbes in your compost bin use enzymes to break down complex molecules into simpler molecules, like water and carbon dioxide. Might some microbes that survive on fallen leaves or the bones of dead animals be able to recycle plastics instead?

To find out more I meet Dr Bronwyn Campbell, a postdoctoral fellow with the Environomics Future Science Platform, a research group set up to use genetics to solve environmental problems.

'The amount of plastic waste that went to landfill in 2018 was more than the combined weight of everybody in Australia,' Bronwyn says. 'It would be great if we could use plastic-munching microbes to degrade that waste.'

Most of the microbes known to degrade plastics are bacteria, but several fungal species can break down plastics too. One of the challenges is that plastic waste is made up of many different kinds of plastics, plus dyes and chemicals present in plastic, food scraps and other waste. Bronwyn's plan is for a community of plastic-degrading microbes that will work together as a team to break down different components of the waste.

'This will be tricky because microbes have a terrible habit of murdering and consuming each other,' she says.

The first step is bioprospecting.

'I want to hunt for communities of microbes in environments where there's already a lot of plastic, like landfills', Bronwyn says. 'I'll also collect microbes where there are hydrocarbons, such as petroleum wastewater facilities. Hydrocarbons are the parent material of traditional plastics and have similar components to plastics.'

Back in the laboratory, Bronwyn will test whether the microbial communities can degrade mixed plastics, then augment them with plastic-munching microbes from international culture collections. The next step will be to work out how to feed the end products back into plastic production, thus transforming plastic waste into a circular resource. It's a promising idea, but Bronwyn emphasises the most important thing we can all do is reduce the amount of plastic waste we make and use.

The desire to care for and conserve the natural world is something that drives many scientists who work at collections. Too often, they are reminded how much has been lost.

5
Lost in time

The word obsoletus *means 'extinct' in Latin. It's the species epithet of* Carcharhinus obsoletus, *the Lost Shark, a species that was described in 2019 but hasn't been seen since 1934. For many species lost in time, museums and collections are the only record. Others – like the Loch Ness monster – are 'lost' as they simply never existed to begin with.*

Paradise lost

A pair of unusual parrot specimens lies inside a drawer in the vault of the Australian National Wildlife Collection (Plate 10). Their feathers are a mix of light brown, sky blue, pale green and red. Unlike other bird specimens in the collection, their feet are posed as though to balance on a perch, they have glass eyes and there is no data on their labels about when or where they were found. An old photograph lying next to them reveals why. Theses parrots are not conventionally prepared scientific specimens. They were once part of a parlour dome, an ornamental arrangement of flora and fauna displayed under a glass dome. Parlour domes were popular in the Victorian era, when it was fashionable to display nature in the home. Similar glass domes were used for displaying dried flowers or ornaments.

The two parrots represent the Paradise Parrot (*Psephotellus pulcherrimus*). It is officially listed as extinct on the IUCN Red List, which records the last confirmed sighting in the wild as

1928 and lists the causes of extinction as drought and overgrazing.

I'm in the middle of writing the paragraphs above when my mobile phone rings. It's Dr Don Sands, an honorary fellow at CSIRO. Don is a doctor twice over, holding both the Doctor of Philosophy (PhD) he was awarded at the start of his career and the Doctor of Science (DSc) he completed in his 80s. In one of life's strange coincidences, Don wants to talk about Paradise Parrots and the moths that live with parrots in termite mounds.

The two closest living relatives of the Paradise Parrot – the endangered Golden-shouldered Parrot (*Psephotellus chrysopterygius*) and the Hooded Parrot (*Psephotellus dissimilis*) – nest in termite mounds. Each has an associated moth species that also lives in termite mounds. The moth larvae are coprophagous, meaning poo-eating. They keep the birds' nests clean by eating the parrot chicks' poo.

When the chicks leave home, the moth larvae spin silken cocoons. Don tells me they remain in a pre-pupal diapause until a cue in nature triggers them to pupate and emerge as adult moths, timed for the parrots' next breeding season.

'I believe that the parrots and the moths have coevolved over millions of years,' he says.

Back in 2007, the late Ted Edwards from the Australian National Insect Collection led research describing the moth that lives with the Hooded Parrot as a separate species from the moth that lives with the Golden-shouldered Parrot. Don recalls helping Ted rear moths from cocoons collected from termite mounds where Hooded Parrots nested. They lodged the resulting adult moths as reference specimens in the insect collection, so the species could be described. Don has phoned me today because he'd like to spread the word about these parrots and their moths to ensure their survival.

'The Golden-shouldered Parrot's moth pupates in the outer walls of the termite mounds,' Don says. 'This is different from the Hooded Parrot's moth, which pupates in soil or loose matter at the base of the termite mounds. I think this makes the Golden-shouldered Parrot's moth more exposed and vulnerable to fire, although I don't know this for sure.'

The Paradise Parrot also nested in termite mounds, digging an entrance tunnel and creating a nest hollow for its eggs and chicks. Don suspects the Paradise Parrot would also have had its own unique moth species living in its nest and keeping its chicks clean. Ted's paper on the Hooded Parrot's moth notes that he tried to find out if that was true: '*A search of museum collections has failed to locate a preserved mound and nest hollow of the paradise parrot which could be checked for silken tunnels and cocoons.*' The Paradise Parrot's mystery moth is presumed to have gone extinct with the parrot.

The search for a Paradise Parrot nest in the world's museums and collections was unsuccessful, but collections do indeed contain bird nests. The most impressive in the wildlife collection is held in a box measuring 85 × 65 cm. It takes up all of the available space. It's the strange construction of a magpie, with a small, soft bowl for the eggs and chicks surrounded by an enormous tangle of coat hangers, fencing wire, computer cables, a phone charger, fabric cord and a pair of 3D glasses from a movie theatre. One of the earliest records of this kind of nest-building behaviour is a paper published in the scientific journal *The Emu* in 1903. It tells of magpies in Tasmania that built nests from the wire used by farmers to tie wheat into bundles at harvest time. When the farmers switched to using string, the magpies switched back to using natural materials for their nests.

Not far from the drawer with the Paradise Parrot specimens and the box containing the magpie's nest is another unusual specimen. It's the skull of a Thylacine (*Thylacinus cynocephalus*; Plate 11), Australia's infamous extinct animal. The species was hunted to extinction in Tasmania by 1936, but this skull is not from Tasmania. It's thousands of years old and was found in 1969 in a cave near Canberra. The Thylacine used to live across mainland Australia, where it went extinct during the mid-Holocene.

On one of my forays into the wildlife collection, I run into Dr Pietro Viacava. His role involves digitising the collections and today he is sorting out boxes containing very old microscope slides. He holds one up to the light. It looks like a purple sliver of tissue on a small strip of glass. The slide is one of a set showing sections of the brain of an adult Thylacine that died in Berlin Zoo in 1880. These slides are nearly 150 years old and have a long history of storage in various collections in Germany before they were lodged here in the wildlife collection. I had seen the news headlines about a week earlier. Researchers at the University of Queensland (UQ) used these slides to upload high-resolution microscopy pictures online and compare the brain of a Thylacine to the brains of wolves and other mammals.

Although it had stripes like a tiger and a body shape like a wolf, the Thylacine was not related to either. It was a marsupial, more closely related to a kangaroo or a koala than to a tiger or a wolf. Any similarity is either a coincidence or an example of convergent evolution, where unrelated species living similar lifestyles in similar habitats come to resemble one another, like echidnas and hedgehogs. The UQ researchers found that rather than resembling the brains of wolves, the Thylacine's brain resembled those of carnivorous marsupials to which it is most closely related, such as Tasmanian Devils and quolls. This suggests that body similarities do not always imply similar brain structures.

Extinct insects

I drop in to ask fly researcher Dr Keith Bayless whether he has any experience working with extinct species in the insect collection. He says most papers describing new species use old specimens from museums and collections as part of the process of comparing the new species to existing ones.

'You might have heard the phrase "rare things are common and common things are rare",' he says.

I nod, although I'm thinking this is actually the opposite of what I've heard: common things happen commonly. I catch up a few hours later when I realise I see galahs and magpies daily, but a long list of other birds rarely. It's rare for a species to be common, and most species are rare.

'Up to 30 per cent of species are described from a single specimen and an even larger chunk are described from specimens found on a single collecting trip,' Keith says. 'That's because you might find a cluster of the same species in the same place in the same conditions, which are difficult to reproduce.'

I ask Keith whether this only applies to insects and he smiles, like he's letting me in on a secret.

'Most species *are* insects,' he says.

I recall someone telling me that most insects are beetles, but I decide it would be polite to not mention this to a fly researcher. Our conversation meanders to Malaise traps, those tent-like contraptions that channel flying insects into a bottle of ethanol. Keith says there have been many times when he has caught an insect in a Malaise trap and then been unable to catch a second specimen until he happens to set up exactly the right kind of trap.

He mentions a skipper fly from Northern Europe, *Thyreophora cynophila*, that went missing. It was a carrion eater, known to live inside the carcasses of large mammals like cows. It's rare now to find a large mammal left to rot in populous Europe, and the fly wasn't seen between 1850 and 2009. The fly didn't go extinct, it simply changed its range and habitat, moving south and switching to feeding on buried carcasses. Some researchers finally caught up with it in Spain when they used a buried ball of squid as bait.

'This particular fly has an orange head and a blue body,' Keith says, bringing up some images on his computer screen. 'Some people think its head is bioluminescent. But ...' he shakes his head.

'Wishful thinking,' I say, thinking of a Facebook group I'm familiar with where people go mad for glowing fungi and sea sparkle, a species of algae that we have in the Australia National Algae Culture Collection.

'There are true bioluminescent fly larvae, the cave glowworms in Victoria and New Zealand,' he says.

With a few more keystrokes he finds a related Australian fly that also has an orange head, *Piophilosoma*. He points out two sharp spines on the male's back.

'They use these spines during mating. If another male fly comes along and tries to knock them off the female, they shank it.'

We both laugh. There are endless strange stories in the world of flies, but Keith makes a rare departure to remind me of Wallace's Giant Bee (*Megachile pluto*). This species lives only on a small number of Indonesian islands and is collected and traded illegally, despite being listed as vulnerable. There's one specimen here in the insect collection, which was intercepted at Australia's borders by biosecurity officials in 2022.

Wallace's Giant Bee is the largest bee in the world. Females have a wingspan over 6 cm, making them a prize for collectors. The common name acknowledges Alfred Russell Wallace, who collected the first scientific specimen of its kind in 1858 and worked with Charles Darwin on the theory of evolution. Curiously, this bee is another species that lives in termite mounds, taking advantage of the protection of the termite colony. They can sting, but unlike European Honey Bees, they don't die afterwards. Sadly, Wallace's Giant Bees don't make giant pots of honey. Instead, they specialise in producing a kind of resin to protect their hives.

Down the hallway from Keith, Dr Juanita Rodriguez recently described a new species that is long extinct. It's a fossil sawfly (Plate 12) from McGraths Flat, the same site that yielded the fossil caddisflies we met in Chapter 4. Like the caddisflies, it dates from around 11–16 million years ago.

Sawflies resemble caterpillars when they are larvae and flies when they are adults, but they are actually an ancient family of wasps that lack the typical 'wasp waist'. In Australia, spitfires are the most widely recognised group of sawflies. Juanita tells me many Australian sawfly species feed on the leaves of Myrtaceae, a family of woody plants that includes eucalypts.

'Many sawflies have anatomical adaptations that enable them to separate out the toxic oils in myrtaceous leaves,' she says. 'Spitfires spit the oils out and use this as a defensive strategy.'

The sawfly from McGraths Flat is the first fossil sawfly from Australia and only the second in the world. It is so well preserved that Juanita was able to describe it as a species. She then fitted it into the sawfly family tree, which enabled her to date the tree.

'We found that sawflies originated in the Cretaceous Period, around 100 million years ago,' she says. 'These ancient ancestors lived in Gondwana. Today all sawflies live in Australia and South America, except one species that lives in North America.'

Pollen grains on the fossil sawfly's head reveal it had recently visited a flowering *Quintinia* plant. Dr Michael Frese, who also analysed the caddisfly fossils, said this provided further evidence for complex species interactions in the environment at McGraths Flat. Collecting pollen and spores enabled him to date the fossil site.

'Plant and animal fossils indicate that McGraths Flat was a pocket of rainforest surrounded by a drier environment,' he says. Finding the right pollen and spores enabled him to determine when McGraths Flat was deposited. 'We think that the fossils from McGraths Flat are 11–16 million years old.'

Neither here nor there

Mention cryptids in CSIRO's collections and the staff will probably assume you mean undescribed species. They might pull out a drawer to show you some unnamed species of mosquitoes or start talking about 'dark taxa', mysterious DNA sequences that exist in nature but can't be matched to any known species.

Mention cryptids in other circles and people will probably think you're talking about creatures like the Loch Ness Monster or the Kraken, creatures that exist only in the human imagination. Or do they? In 2013 Sir David Attenborough told the audience at a media event in the United Kingdom that he thinks there might be something to stories of the Yeti. And one day at work I mention the Loch Ness Monster, only to discover that one of my colleagues, Dr Mike Hodda, once worked on Nessie. It's almost as shocking as the day I learnt the herbarium is haunted, though I suspect the 'ghost' is just a creepy feeling people get from sitting with their backs to dark rows of shelving, like being in a library at night.

Mike works at the insect collection on nematodes, which of course are not insects. They are the tiny roundworms we met in Chapter 1. Mike studies nematodes that are biosecurity pests, like the Pine Wood Nematode we will meet in Chapter 7. If anyone could find something the size of the Loch Ness Monster, surely it would be a scientist who can find tiny nematodes. I ask for details.

'I worked on Project Urquhart, which was a scientific exploration of the fauna of Loch Ness in the 1990s by the Natural History Museum in London,' Mike says. 'Castle Urquhart is the ruined castle on the bank of the loch.'

Of course, the project didn't reveal *Nessiteras rhombopteryx* – the species name given to the Loch Ness Monster by the two scientists who named it (just in case) in a paper published by the prestigious journal *Nature* in 1975. If there's no monster in the loch, what did they put a name on? Circus elephants taking a bath? A plesiosaur left over from the age of the dinosaurs? An overturned rowing boat?

'Some of the theories for Nessie do not have a lot of believability,' Mike says. 'There is almost no shallow water around the shores of the loch for elephants to stand in, which I know from trying to sample nematodes without a boat.'

'I do have a fuzzy, underexposed picture of a strange, dark object emerging from the waters of Loch Ness – which I can confirm is me in a dry suit because the water is very cold. There was a Discovery Channel documentary on Project Urquhart, and after it aired someone nominated me for a bravery award for going in the loch despite the presence of the monster! Needless to say that did not go anywhere, but we did find new species of nematodes on the bottom of the loch which are left over from the Jurassic, are predaceous and look pretty scary just like Nessie. The only thing is that they are about a million times smaller than Nessie is supposed to be.'

In 2019, researchers from New Zealand visited Loch Ness to use environmental DNA (eDNA) to search for an answer to the mystery. eDNA is DNA shed by organisms into the environment, enabling species to be detected forensically by sequencing eDNA in samples of water, soil, sand or even air. The search for fish and aquatic animals in the loch revealed a large amount of eel DNA in the water. But a question does linger. Using eDNA to detect species depends on being able to match eDNA sequences from environmental samples to reference sequences of known species. If you're looking for the eDNA of a species unknown to science, how do you know what to look for?

Living giants of the deep

Loch Ness hasn't given up its monster, but the ocean has revealed creatures akin to the Kraken, albeit far too small to sink a ship. It was only in 2012 that a Giant Squid was filmed in its natural habitat for the first time. Since then, I've longed to see one, which is why I find myself unspeakably excited to visit the collections of Tasmanian Museum and Gallery (TMAG) on the day that staff are thawing a squid in a large tub on the paved walkway outside their laboratories. It's not a Giant Squid, but it is a specimen of a large species called *Taningia danae*, the Dana Octopus Squid.

The smell on the walkway is strong, mainly coming from smaller tubs where specimens are macerating in water. This process lets bacteria clean bones of soft tissue before they are stored in skeleton collections. Our wildlife collection mainly uses hide beetles, flesh-eating beetles that are stored a long way from the collections building. Specimens are then frozen to kill pests before they enter the wildlife vault.

Our guide lifts the lid of the large tub to reveal a mottled purple and white squid around 2 m long, folded back on itself. This specimen has been stored frozen in a block of ice for two decades since it was caught and donated to TMAG. By the time we arrive, it is floating in icy water and has thawed sufficiently for our guide to move it around. He bends over the tub and lifts one of the long feeding tentacles to show us a photophore, a structure on the end of each tentacle that can open and close like an eye to emit light, presumably to attract, illuminate or disorientate prey. Next, he prises apart some folds of tissue to reveal the squid's sharp beak, which looks like it could snap off a finger. Soon, the staff will replace the icy water with 10 per cent formalin to preserve the squid's tissues then move it into 75 per cent ethanol for long-term storage.

Later, I look up the anatomy of a squid to make sense of what I've seen. A squid is a kind of mollusc, like a snail or an oyster, but it bears little resemblance to either. A typical squid has a cone-shaped mantle with a pointed tail, short fins and a head with large eyes, below which eight arms and two tentacles extend. Most species have suckers along the tentacles. I look back at the photos I took and notice prominent suckers along the arms of the Dana Octopus Squid.

I flick through my other photos from TMAG: the skull of a sperm whale on a supporting rig in the driveway; the skull of a Tasmanian Devil, its left side eaten away by facial tumour disease, leaving greyish bone with sponge-like holes; the skull of a seal and a section of its fur, branded long ago during its life with a tracking number; and a drawer of Australian moths with a white metallic sheen and different patterns of light brown lines that may or may not indicate they are different species.

It strikes me that despite the scepticism cryptids generate, there are thousands of unusual animals in the real world. Perhaps some cryptids have sprung from deep cultural memories of long-extinct species.

Giant birds and a lost shark

Madagascar's elephant birds were the largest birds to ever live, reaching up to 3 m tall and weighing more than 500 kg. They died out around 1,000 years ago, but pieces of their ginormous eggs are still abundant on the island's beaches.

Dr Alicia Grealy works with the collections at CSIRO, where she has an almost legendary status for her skills in working with DNA. In one instance, she extracted DNA from a putative Paradise Parrot egg allegedly collected 40 years after they went extinct. Her work proved it was just a Bluebonnet, a common native parrot.

During her PhD at Curtin University, she studied ancient DNA from the eggshells of elephant birds. DNA doesn't survive well in hot, humid climates like Madagascar, but the thick bioceramic of elephant birds' eggshells has protected their DNA across millennia.

'Because skeletal material is scarce, eggshells offer a unique opportunity to study these birds to learn more about the origins of biodiversity and about extinction,' Alicia says. 'The island of Madagascar itself is also important for understanding evolutionary processes.'

Elephant birds are related to emus and cassowaries, but their closest living relatives are the much smaller kiwis of New Zealand, something that was already known from ancient DNA. Alicia wanted to go deeper to find out how many different species of elephant birds existed. Her PhD involved examining 960 elephant bird eggshell fragments ranging in age from 1,300 to 6,200 years, and two years of painstaking laboratory work. The result was a family tree of elephant birds.

'At the time of their extinction there were likely three species belonging to two different families,' Alicia says. Although skeletal morphology had suggested there were more species, we think this was due to extreme differences between males and females of the same species.'

Ancient DNA also provided Alicia with clues about how the past environmental conditions in Madagascar shaped the evolution of elephant birds.

'We used the amount of genetic difference between species to estimate when they split from a common ancestor,' Alicia says. 'For the two families, this happened around 30 million years ago. At this time, populations of small elephant birds may have adapted to a changing climate, eventually becoming different species, with some remaining small and others becoming quite large.'

'More recently, a second split happened among the larger of the elephant birds, which coincided with another period of climate change around 1.5 million years ago, leading to the evolution of an even larger species. Between this time and their extinction, one species, *Aepyornis maximus*, doubled in size. This shows extreme gigantism can evolve over very short timescales, a surprising result.'

Alicia's team also measured stable isotopes from the eggshells, which revealed elephant birds' diets were a mixture of shrubs, succulents and grasses.

'We found elephant bird species had different diets, suggesting they adapted to unique ecological niches. Changes in the environment, including vegetation, may have driven speciation.'

Elephant birds have left evidence of their existence lying on beaches a millennia after their extinction. Other lost species have left fewer traces.

Carcharhinus obsoletus, the Lost Shark, was scientifically named in 2019 by a team of researchers including Dr Will White from the Australian National Fish Collection. The species is a whaler shark in the family Carcharhinidae, an economically important family that includes dozens of familiar species like the Bull Shark and the Lemon Shark.

Will was aware of a whaler shark from the South China Sea that was likely to be an undescribed species. He was keen to understand its conservation status in this heavily fished part of the world. Sadly, the team found only three specimens in the world's museums and collections, which were collected from Borneo, Thailand and Vietnam. They are held in the Academy of Natural Sciences in Philadelphia and the Naturhistorisches Museum in Vienna.

The most recently collected was the specimen from Vietnam, dating from 1934. The species hasn't been recorded since, despite many biodiversity surveys at sea and shark surveys at fish markets.

Will and the team compared the specimens with closely related shark species in the family, studying features like body measurements, teeth shape, skeletal X-rays and microscopic features of skin structure. Their paper describing the Lost Shark reports in detail on characteristics like *'pronounced dignathic heterodonty between upper and lower jaws', 'pectoral fins short and relatively broad'* and *'tip of ventral caudal-fin lobe narrowly rounded'*. A painting by artist Dr Lindsay Marshall, whom the shark *Heterodontus marshallae* that we met in Chapter 2 was named after, imagines the holotype of the Lost Shark as a living shark (Plate 13). It reveals a pale grey shark with a slender body, small dorsal fins, a short snout and large eyes. It probably grew to around 1 m in length.

The Lost Shark likely lived in shallow waters less than 50 m deep and likely became a victim of overfishing. The IUCN Red List records the Lost Shark as critically endangered, with fewer than 50 individuals left in the wild. But with no sightings in nearly 100 years, the species may live up to its name.

The mystery surrounding lost creatures piques our curiosity and should remind us to care for the biodiversity that surrounds us, in all its weird and wonderful forms.

6
Curiouser and curiouser

In the worlds of insects and spiders, nature has invented many strange creatures. If they were scaled up to the size of a cat or dog, they would be both extraordinary and terrifying. But there are even stranger things in the sea, from the purses of mermaids to curious creatures inhabiting the perpetually dark waters of the abyss.

Dung rollers

Ancient Egyptians revered the dung beetle *Scarabaeus sacer*, the Sacred Scarab. It featured in their jewellery, seals and amulets. Khepri, the god of the morning sun, had the head of a scarab. The motif is still a popular in jewellery today, despite dung beetles performing one of lowliest tasks in nature: recycling poo.

Dung beetles feed on poo and lay their eggs in it as a source of food for their larvae. Species include 'rollers', which collect poo by rolling it along the ground like the Sacred Scarab; 'tunnellers', which bury poo in the ground beneath where it lies; and 'dwellers', which simply live in poo wherever they happen to find it.

The scarab beetle family, Scarabaeidae, has more than 35,000 species. As well as dung beetles, the family includes many other species that feed on decaying animals and plants. Around the world you'll find rhinoceros beetles with long horns, bright gold scarabs with an almost mirror finish and Australia's Christmas beetles.

Australia has more than 500 native dung beetle species. They evolved to clean up the small, dry, fibrous poo of native marsupials. Sloppy cow pats aren't on their menu, which meant that the introduction of cattle farming in Australia created a very stinky problem. Cows defecate 10–12 times a day, producing enormous numbers of pats that can take months or even years to break down. But the problem was not just about the stink. These pats became an abundant breeding ground for native Bush Flies and midges, which boomed in numbers and became pests.

In 1955, Hungarian-born scientist Dr George Bornemissza OAM joined CSIRO as a research scientist. He soon made the connection between the cow pats littering paddocks and Australia's fly problem. Back home in Europe, cow pats tended to be cleaned up quickly due to the presence of suitable dung beetle species.

Bornemissza spent years working on a project that would convince CSIRO to import dung beetles that were equal to the task of eating cow pats, including nine years in Africa selecting and breeding suitable species. In 1967, CSIRO began importing what would become dozens of dung beetle species from Africa and the Mediterranean. The dung beetles improved pastures and helped reduce Australia's fly problem.

The Australian National Insect Collection still holds Bornemissza's display drawers of beetles. It was a project he continued into retirement, travelling the world, visiting museums and acquiring specimens to show the great diversity of the world's beetles. His drawers are part artwork, part scientific collection. Some show specimens of related species in an array of colours and patterns arranged in clockwise circles from largest to smallest. The last specimen in the circle is about the size of the knee of the first.

Almost all dung-rolling beetles belong to the scarab family, but Australia has some dung-rolling species with a difference – they are weevils. It's an example of convergent evolution, where unrelated species evolve similar physical traits or behaviours, like the spines of echidnas and hedgehogs or the body shape of sharks and dolphins.

There are close to 200,000 weevil species in the world, just over 62,000 of which have scientific names. They are among the cutest of beetles, easy to recognise thanks to their long snouts. In most species, the snout – or rostrum, to be more scientific – is used to drill into plant tissue to feed or lay their eggs. Unlike most weevils, dung-rolling weevil species feed on poo. In a way, they do eat plants – but after the plant has been processed through the gut of a marsupial.

Dr Hermes Escalona of the insect collection scientifically documented the dung-rolling behaviour in 2022, filming it at Undara Volcanic National Park, an area in north Queensland renowned for its lava tubes (Plate 14). The poo-rolling weevils belong to the genus *Tentegia*.

'Weevils of the *Tentegia* clan roll dung pellets of marsupials into collections under logs,' Hermes says. 'The females lay their eggs in the pellets and the larvae eat them as they develop.'

Tentegia weevils live in monsoonal and arid parts of Australia. Several species seem to be quite rare, living only in small areas.

'The behaviour of rolling and eating dung is unique in insects outside of the dung beetle realm. These weevils have come to resemble dung beetles as part of their adaptation to arid Australian ecosystems.'

Together with colleagues Dr Rolf Oberprieler and Debbie Jennings at the insect collection, Hermes studied the taxonomy of the dung-rolling weevils. They gave scientific names to three new species of *Tentegia* weevils from the Northern Territory (NT). This brought the tally of *Tentegia* species in the NT to eight, more than anywhere else in Australia. One of them lives on CSIRO's site in Darwin.

Mermaids' purses

In Chapter 2 we met *Heterodontus marshallae*, the Painted Hornshark, a recently named shark species that swims in the waters off the north-west coast of Western Australia. It's an oviparous shark, meaning it lays eggs.

After baby sharks hatch, the strange objects that nurtured them as embryos sometimes wash up on beaches. They're often confused for things like seed pods or other marine creatures. In more imaginative circles, they're thought to be mermaids' purses. Between 4–25 cm long depending on the species, they're just the right size to hold a comb and some lip gloss (Plate 15).

The egg cases of oviparous chondrichthyans – the formal way of saying egg-laying sharks, rays and chimaeras (which are relatives of sharks and rays sometimes called ghost sharks) – are made mostly from the protein collagen. They come in a huge variety of colours, from butterscotch through amber to very dark brown. Their shapes couldn't be further from egg-shaped, ranging from flattened rectangles to corkscrews. Some are smooth, others have ridges and keels. Many have 'horns' at each corner or long, winding tendrils like a vine.

While working on the description of the Painted Hornshark, Helen O'Neill from the Australian National Fish

Collection was involved in the search among museums and collections for specimens of the unnamed shark. The search unearthed one egg case. It's dark brown, corkscrew-shaped and has long, curled tendrils emerging from the point of the screw.

The closest relative of the Painted Hornshark is the Port Jackson Shark. It also lays dark brown, corkscrew-shaped egg cases, which are more tightly spiralled.

'The corkscrew shape helps the egg wedge in rock crevices so it doesn't get swept away with the tide,' Helen says. 'The vine-like tendrils of many kinds of egg cases are attached to coral or kelp to stop them drifting away. The mother does this by hooking the tendrils onto an object and swimming around and around during laying.'

The incubation time for a chondrichthyan embryo in an egg case ranges from a few months to three years, depending on the species. Helen tells me that by the time an egg case washes up on a beach to be mistaken for a mermaid's purse, the embryo has likely already hatched or died. Some embryos end up eaten by creatures like sea snails, innocent-looking shelled creatures that bore holes through egg cases and suck out the contents.

Not all sharks lay eggs. Some give birth to live young, but there are many variations on the theme. Some species have a placenta-like structure, meaning the embryos receive nourishment from the mother's body. Others develop within eggs within the mother and are nourished by a yolk, before hatching in the womb and then being born. Want more variety? Say hello to multiple paternity, where the litter has more than one father; parthenogenesis, where the litter has no father at all; oophagy, where baby sharks eat unfertilised eggs in the womb; and finally adelphophagy, where baby sharks eat each other in the womb.

The egg cases of oviparous chondrichthyans are still something of a scientific mystery. Each species' egg case is unique, but who laid which egg case? It's like a game of pairs for taxonomists, where the goal is to match each kind of egg case to the species that lays it. Helen's role in the game includes searching inside shark specimens in the fish collection.

'We borrow egg cases from other collections, museums and aquariums around the world,' she says. 'We also have our own specimens collected from fish markets, surveys at sea or extracted from the reproductive tract of preserved specimens in our collection.'

The work is supported by citizen scientists who are part of a long-running project to record sightings of egg cases at beaches around the world. Known as the Great Eggcase Hunt, the project is run by a charity, the Shark Trust. It began in the United Kingdom more than 20 years ago and people have recorded hundreds of thousands of sightings of egg cases.

The Great Eggcase Hunt launched in Australia in 2023, in partnership with CSIRO. Citizen scientists logged more than 2,400 sightings of egg cases in Australia in the first 18 months. There are identification posters available online and through the Shark Trust mobile phone app if you enjoy beachcombing and would like to get involved.

'Matching egg cases to species is important for understanding the basic biology of oviparous chondrichthyans,' Helen says. 'Egg cases can act as a calling card, revealing where different species occur and where their nurseries are located. If a lot of egg cases wash ashore in one place, that's a clue an important breeding ground is nearby.'

Resident weevil

In 2021, a weevil took up residence in trees in Perth, Western Australia. The Polyphagous Shot-hole Borer (PSHB; *Euwallacea fornicatus*) measures just 2 mm long and has a curious way of reproducing that means a single female can give rise to a new population with no need for a male.

In more familiar sex-determination systems, males and females have two copies of each chromosome – they are diploid. Eggs and sperm contain only one copy of each chromosome – they are haploid. Offspring inherit one set of chromosomes from each parent. One of those pairs are sex chromosomes, giving rise to XX females and XY males in humans, ZZ males and ZW females in birds, and so on. There are endless variations in nature, including dragons whose sex chromosomes can be overridden by the temperature at which their eggs incubate.

In the PSHB, males develop from unfertilised eggs and thus have only one copy of each chromosome – they are haploid. They tend to be smaller than females and don't usually fly. Females develop from fertilised eggs, and thus have two copies of each chromosome – they are diploid. Using a haplodiploid sex-determination system means that a single female can lay eggs that will develop as males. By breeding with those males, she can then produce females, which then breed with their male siblings. In this way a single female can give rise to a founding, invasive population.

Dr James Bickerstaff is a postdoctoral research fellow at the insect collection. He works on weevils like the PSHB.

He says individuals of this species usually mate with their siblings, which is actually a good thing in this case.

'The resulting inbred offspring tend to be healthier than outbred offspring,' he says. 'It's pretty weird but it explains why they're successful invaders.'

James goes on to explain that inbreeding and haplodiploidy could help populations of PSHB get rid of harmful mutations swiftly. In diploid species, a harmful mutation can be maintained in a population if an individual can get by with a normal copy of the gene on one chromosome and a harmful copy on the other. But in PSHB males, there's no second copy to compensate for a harmful mutation and any male carrying one may die before it can reproduce. In a similar way, a mutation that confers some kind of advantage can quickly become fixed in the population.

The PSHB is part of a group of weevils called ambrosia beetles. Named for the food of the Greek gods, ambrosia beetles tunnel into trees and farm fungi inside those tunnels. Although it occurs among closely related weevil species, fungi farming has evolved independently at least 16 times. In the case of the PHSB, trees can be overwhelmed by the number of weevils tunnelling through them or killed by the fungal strains they carry.

Ambrosia beetles have lost the cute, nose-like rostrums that are characteristic of weevils. This is probably because a long nose would get in the way of tunnelling through wood, which is obviously much tougher than nosing into soft stems or poo. A curious feature they share is the wide variety of bodily structures to store the fungi strains they farm.

'Ambrosia beetles have specialised organs called mycangia,' James tells me. 'They use these to transport their fungus from the tunnels where they are born to the new homes they create.'

'The morphology of mycangia vary widely. Some are sack-like organs inside their head capsule. Others are pits in their exoskeleton.'

Most ambrosia beetles tunnel into dead wood. But when they invade new places, they tend to attack living trees. The PSHB can attack hundreds of different tree species. With no treatments available, the best option is to remove infected trees to stop the spread.

James recently led a project to sequence the genome of the PSHB as the starting point for understanding the genetics and physiology of the species, with a view to developing treatments for infected trees.

Comparing its genome with a related native weevil species revealed the PSHB is different in the way its genes are arranged and it has many sections of repeating DNA sequences.

Among the many other weevils resident in the insect collection is a tiny specimen collected in 1836 near present-day Albany in Western Australia. It was collected by English scientist Charles Darwin during his voyage on HMS *Beagle*. The weevil is a dark, mottled brown colour and far too small to be pinned through the body, as is usual. Instead, it is glued to a sliver of cardboard that is pinned through the weevil's labels into the foam lining its tray. This tiny specimen is not much bigger than the head of the pin securing it.

Darwin's weevil is far from the oldest specimen in the collection, losing that competition by around 100 million years to weevils fossilised in amber during the Cretaceous. They are preserved so beautifully they look like they fell into a drop of honey just a few minutes ago.

Amber is the fossilised resin of trees, which were probably pines in the family Araucariaceae. Today, these kinds of pines are less widespread than during the Cretaceous and include rare Australian trees like the Wollemi Pine and the Queensland Kauri Pine.

While de-extinction of dinosaurs *Jurassic Park* style remains science fiction (see Chapter 8), insects preserved in amber can help researchers paint a scientific picture of the past. Weevils in amber have been used by staff at the insect collection to help construct phylogenetic trees, understand how insects and flowering plants evolved together, and even name dozens of long-extinct species of weevils that were alive during the Cretaceous.

Jumping fossils

As its name suggests, the Australian National Insect Collection specialises in specimens with six legs – it has around 12 million of them. But it does make a few exceptions to the rule, including many-legged millipedes, legless nematodes and eight-legged members of the family Salticidae, also known as jumping spiders. These are known for their large eyes, ability to jump long distances relative to their tiny size and the peacock-like mating displays of some species. Australia has around 1,200–1,500 species of jumping spiders, but only 500 or so have scientific names.

Dr Barry Richardson is an honorary fellow who researches jumping spiders. One day, he happened to be working beside Dr Michael Frese, whom we met in Chapter 4, at the insect collection's photomicroscopes.

'Michael noticed a jumping spider on my screen and offered to show me a tiny spider fossil,' Barry said. 'I knew straight away we were looking at the first fossil jumping spider from the southern hemisphere.'

Fossil jumping spiders are rare and most have been found in Baltic or Mexican amber. But, like the caddisflies we met in Chapter 4 and the sawflies we met in Chapter 5, this jumping spider fossil was found at McGraths Flat, the fossil deposit in NSW that is known for its exceptionally well preserved fossils from the middle of the Miocene Epoch, 11–16 million years ago.

Barry shows me photographs and scanning electron micrographs (SEMs) that depict the spider from its dorsal, or top, side. It's 2.5 mm long with a short, rounded cephalothorax (head and thorax). The large lenses of its two front eyes are surrounded by short, stiff hairs called setae. It still has one fang and most of its legs, complete with joints and setae. Some of the spider's internal structures have been preserved, including parts of its brain and pharyngeal plate, which sieves food inside the mouth.

To ensure that finding details of the mouthparts wasn't too good to be true, Barry dissected modern jumping spider specimens held in the insect collection. Michael then gold-plated their mouthparts and took more SEMs. (Non-conductive materials need a thin metal coating for SEM, hence the gold layer. This step is not necessary for taking SEMs of fossils from McGraths Flat because the rocks are iron-rich.)

'Michael found that the pharyngeal plate of the fossil is quite similar to the modern specimens,' Barry says. 'He also realised there was a strange set of parallel tubes that appeared to be a neuropile, a bundle of nerve axons connecting one of the eyes to the brain. Their diameter would have affected the speed of signal transmission along the nerves. Having fossils that allow us to study the evolution of brain functions is very exciting.'

Barry and Michael put their heads together with Dr Matthew McCurry, a palaeontologist at the Australian Museum, to take things a step further. They put together multiple sources of information, including the fossil jumping spider's morphology; a map of where similar spiders live today, which was created using specimens held in museums and collections; and the past climate, based on fossilised vegetation at the site. They also identified the fossil spider's genus, realised that its close relatives are still alive today and mapped where to find them.

'The fossilised jumping spider belonged to *Simaetha*, a genus of jumping spiders that still exists,' Barry says. 'From DNA studies, we knew the group this genus belongs to probably evolved in Australia only a few million years before the time this spider at McGraths Flat died. The fossil shows that the distinctive morphology of *Simaetha* had evolved fairly quickly, but for the next 15 million years it didn't change! We wanted to know why.'

'Using information on the climate profile of the fossil site generated through analyses of the fossilised vegetation, Matthew mapped where such conditions occur today. We then compared this with the predicted distribution map of *Simaetha* today, made using the specimens in museums around Australia.'

'Lo and behold, the two maps matched. Modern species are found in the same habitat and climatic conditions as the fossil, although it is now in eastern Queensland rather than central NSW. Not only do the fossil and modern spiders look the same, they also still live in the same type of environment.'

Deep sea strangers

Some of the strangest-looking animals reside in the deepest parts of the ocean, the abyssal waters. Perhaps there is something about the lack of light and extreme pressure at these depths that leads to animals taking on curious forms. But looks are irrelevant when it's perpetually dark except for flashes of bioluminescence.

The abyss off the east coast of Australia is one of the world's least explored places. In 2017, RV *Investigator* sailed to this region with a simple goal: to carry out the first ever study of what species call it home. It might sound curiosity-driven, which is partly what makes a field trip at sea exciting, but it was also science-driven – biodiversity surveys like this are the source of information for managing marine areas. Like all voyages of RV *Investigator*, 'Sampling the Abyss' was a partnership of many organisations, this one led by Museums Victoria.

Researchers use a variety of techniques to collect specimens from the depths of the ocean, down to 5 km. This includes using a small beam trawl net and a benthic sled, which samples the sea floor. RV *Investigator* has an on-board laboratory where specimens are sorted by species, photographed and measured. Muscle tissue is taken from fish specimens for DNA analysis and then the specimens are preserved or frozen for later preservation. At the end of a voyage, specimens are sent to various museums and collections around the world, where they continue contributing to research over years and even decades.

The scientists sampling the abyss in 2017 found many, many curious creatures during their travels, totalling thousands of specimens. Most of them were invertebrates although fishes were also found.

John Pogonoski, an ichthyologist at the fish collection, was one of the on-board science team tasked with identifying the abyssal fish catches, including a Faceless Cusk (*Typhlonus nasus*), a Blobfish and a Deepsea Lizardfish (*Bathysaurus ferox*).

'I was a bit gobsmacked when I saw the Faceless Cusk for the first time,' John says. 'It took a few hours of hunting through the scientific literature to work it out. It had never been seen in Australian waters, but we now know it rarely occurs above 3,000 m depth, which explained its absence from earlier Australian surveys when we didn't have the equipment to sample that deep.'

It may be no surprise to learn that the Faceless Cusk has very minimalist facial features (Plate 16). Its head looks like a fleshy stump, its eyes are almost invisible and its mouth is low on the underside of its head. The Faceless Cusk is a species of cusk-eel, which are not true eels but do have an eel-like shape.

Blobfish specimens, often affectionately called Mr Blobby, are a family of fish beloved for their floppy flesh and downturned mouths, but they don't look quite like this during life. Deep in the sea, their bodies are under enormous pressure. A Mr Blobby looks very different photographed on land after it has been caught.

The Deepsea Lizardfish is an ambush predator with the looks to prove it – large eyes, a gaping jaw, needle-like teeth and a long, flexible body (Plate 16). Its genus name, *Bathysaurus*, means 'lizard of the depths'. Deepsea Lizardfish have both male and female reproductive organs, which means any two that manage to meet up in the vast depths of the ocean can breed together.

Last up is *Gesaia csiro*, a polychaete worm that resembles a crumbed prawn. It grows to around 30 mm long and was found off the NSW coast around 4.5 km deep. This new species was named in honour of – you guessed it – CSIRO. The researchers who named this creature were from the Australian Museum. According to their paper describing the species, *Gesaia csiro* lives in 'tubes built of fine sand forming colonies on substrate'. It's the sand that the worm has built up around its body that gives it the appearance of being crumbed like a cutlet.

> *Curious creatures and weeds of biosecurity concern regularly arrive at Australia's ports. Detecting them is one challenge. Being able to tell them apart from native species is another.*

7
Space invaders

As soon as our ancestors began to travel across continents and oceans, pests and weeds began to hitch a ride. Collections are one weapon in the war against their continued introduction and spread. They provide expertly identified reference specimens that can be compared against invaders to verify their identity and answer the question: Are they already here?

Tree body problem

Imagine a tiny roundworm only 1 mm long. Magnified under a microscope, it looks like a droplet of water snaking down a window. This is Pine Wood Nematode (*Bursaphelenchus xylophilus*), a species native to North America. It's extremely difficult to tell apart from harmless nematode species that live in Australia.

As its name suggests, Pine Wood Nematode is a problem for pine trees. It causes an infestation known as Pine Wilt Disease, which can kill a pine tree in as little as six weeks. The nematodes can spread rapidly through gardens, forests and plantations thanks to a complex interplay between a trio of species: the nematode, a fungus and a beetle (plus some strains of bacteria).

The nematodes feed on cells inside pine trees, multiplying rapidly into the millions. Their presence and the tree's response prevent water flow, causing the tree to wilt and die. Trees killed by nematodes are attractive to several species of beetles, which breed inside the damaged trees. The nematodes gather in the breeding chambers of the beetles, attach to their bodies and travel with them to new host trees. The beetles also carry blue stain fungi, which can grow inside the tree after it dies and provide extra food for the nematodes until they find another host tree.

Dr Dan Huston is a research scientist at the Australian National Insect Collection. We met him briefly in Chapter 2 when he named a marine parasite after his baby daughter.

'Pine Wood Nematode could enter Australia in shipping containers, wood chips or timber pallets,' he says. 'Although we don't have the same species of beetles here, they could potentially be picked up by a local species, as has happened when Pine Wood Nematode invaded Asia and Europe.'

Together with his colleague Dr Mike Hodda, Dan researched and wrote the National Diagnostic Protocol for Pine Wood Nematode.

'There are lots of harmless native nematodes, many of them without scientific names,' Dan says. 'They are very difficult to tell apart from Pine Wood Nematode. The diagnostic protocol means that whenever there's a suspected biosecurity incursion, the nematode can be quickly identified and dealt with.'

Mike and Dan's work relied on specimens of nematodes held in the insect collection.

'We have a reference collection of hundreds of specimens related to this project,' Dan says. 'They are stored on microscope slides, which lets us zoom in on the details of different species.'

An ancient problem of biblical proportions

It's midsummer in Australia when I first notice small patches of missing fibres where the woollen carpet in my loungeroom meets the skirting boards. I pull the house apart, finding holes chewed through suits, the crumbled remains of a feather duster, tiny silk cases on woollen jumpers and little caterpillars living behind furniture. Looking up, I notice small moths crawling on the ceiling. They are light brown and slightly metallic in appearance.

The moths responsible for the damage are thought to have left Africa with our distant ancestors, presumably in the furs they carried. They even get a mention in the Bible.

> For the moth will eat them up like a garment; the worm will devour them like wool.
>
> Isaiah 51:8

Clothes moths belong to Tineidae, an ancient family of moths. There are two species that chew their way through our carpets, suits and jumpers: the Webbing Clothes Moth (*Tineola bisselliella*) and the Case-making Clothes Moth (*Tinea pellionella*). They are nature's recyclers, originally more at home in the nests of birds where they fed on fungus and feathers. Nature has invented many creatures to clean up waste, from the dung-rolling weevils we met in Chapter 6 to zombie worms that feast on whale bones at the bottom of the ocean.

In an attempt to halt the invasion, I purchase clothes moth traps from a hardware store. They are small cardboard tents lined with a sticky surface impregnated with female pheromones. They fill up with adult male clothes moths, attracted by the scent of the females, and one cricket, unlucky enough to walk across a trap. I show photos of the traps to moth researchers at the Australian National Insect Collection and take heart that I have some of the world's leading taxonomists to advise me. It's a perk of the job to be able to get curious creatures identified, something I also take advantage of when a cockroach shorts out my hot water service. It turns out to be a native kind, with cream stripes along its sides.

Meanwhile, the moth taxonomists tell me that a few weeks in the freezer will kill both the larvae and eggs of clothes moths. I spend months cycling every woollen item that will fit through my freezer, recognising this is the same strategy that will be used on a larger scale to kill pest insects and eggs when insect and wildlife specimens are relocated to our new collections building. The team will use industrial-sized freezers that drop to −35°C, requiring only a week to kill the eggs of pests like carpet beetles, which are a collection manager's nightmare.

I find out more from Tonya Haff, the collection manager of the Australian National Wildlife Collection. She tells me that the best defence against pest insects is rolling pest checks.

'We ensure each bird or mammal specimen is checked for pests once per year,' she says. 'We line our specimen trays with white archival foam or paper so that signs of pest insects are easy to spot. They like to hide in the tight spaces where a specimen is in contact with a tray, so to check for pests we lift each specimen and look for frass, among other signs.'

Frass is insect waste, unfortunately familiar to me at home in the crumbling patches where my carpet used to be.

'If we find pests in a drawer, we can treat the entire cabinet by freezing or fumigation,' Tonya says. 'We colour code our cabinets with magnetic strips to keep track of when we last checked them for pests and whether they received the all-clear.'

Tonya's team also uses sticky traps similar to clothes moth traps, but they don't contain any pheromones. Instead, the traps they use are just small cardboard tents with a sticky floor. They catch crawling insects that happen to walk across them, alerting the team to any pests that might be lurking in the collection vault.

Like the sudden arrival of clothes moth adults in spring and summer, it's common for people to notice dramatic increases in insect numbers at times. When there are suddenly a lot more pesky flies or charismatic dragonflies around, people often get in touch with the insect collection to ask why.

Dr David Yeates, its director, says fluctuations in insect populations are complex but often just seasonal.

'During warm weather, insects tend to be more active because their body temperature depends on the external environment,' David says. 'During spring and summer, many insect species emerge from a winter resting phase to begin their winged, adult life stages. We notice flying adults more than their flightless larval stages.'

Parrots of the world

The birds in the garden that surrounds my house are much more welcome than the moths inside, even if they do steal all the fruit. Cockatoos take the cherries. Silvereyes eat the figs but look cute jumping into the tiny bowl of water I give them to drink. Currawongs gather in the backyard and sing to each other. Australian White Ibis circle on thermal currents high overhead. Blackbirds – the only introduced species in this list – drink from the dog's bowl when he isn't looking. Wattlebirds visit the protea, Crested Pigeons sit on the roof and Australian Magpies hunt for grubs. When the olive tree is in fruit, parrots gather among its branches: Crimson Rosellas, Eastern Rosellas and my personal favourite, the red and green-coloured King Parrots.

Dr Leo Joseph is the director of the Australian National Wildlife Collection. He's been working on birds since his Honours project on rosellas in the early 1980s. Today he's giving a talk on a recent paper he wrote with collaborators from the USA, a country with no surviving endemic parrots, where escaped pets have established invasive populations.

Instead of diving into taxonomy and phylogenetics, Leo presents a slide show about parrots. There are big parrots and small parrots; familiar parrots and rare parrots; blue and green parrots that blend into the strata of rainforests; and camouflaged parrots like Australia's Night Parrot and New Zealand's Kākāpō, which share colours but not a close relationship. Strangest of all is the Palm Cockatoo from Cape York Peninsula and New Guinea, a black parrot with a long crest of black feathers, bright red cheeks, a bill that it never completely closes, a coloured tongue, and a habit among males of using sticks to drum on trees to enhance their visual and vocal displays.

'Where do parrots sit in the bird family tree? Well may you ask,' Leo says. 'The Order Passeriformes includes maybe half of

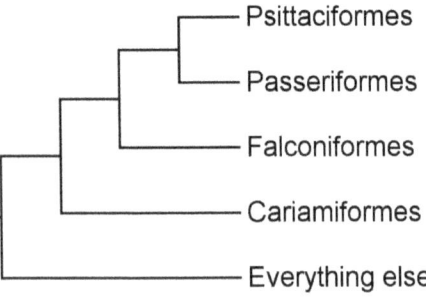

Fig. 1. A phylogenetic tree of birds. Parrots, which make up the order Psittaciformes, are most closely related to Passeriformes, which include the world's songbirds.

all birds. Most closely related to Passeriformes are Psittaciformes, the parrots. The next group out is Falconiformes, then Cariamiformes, then everything else.'

Passeriformes are also known as perching birds. They all have three toes facing forward and one back. They include all songbirds, which arose in Australia before spreading around the world. Falconiformes are, as their name suggests, falcons. Cariamiformes have only two living species, which are long-legged birds known as seriemas that live in South America. Among their long-extinct relatives are South America's terror birds, giant carnivores whose reconstructed skeletons recall birds' dinosaur ancestors. And 'Everything else'? That loosely refers to all other birds – cranes, eagles, ducks, emus, puffins, penguins and more.

Leo and his coauthors' recent paper revised the taxonomy of parrots. They noted that parrots are relatively easy to recognise, thanks to their hooked bills, zygodactylous feet (two toes facing forward and two back) and predominantly green or red plumage. (There are many exceptions, of course, like white and yellow Sulphur-crested Cockatoos in my cherry trees.) But finer taxonomic classification was difficult before DNA analysis allowed their relationships to be better understood.

Leo finishes his talk by reflecting on the colours teal and yellow, the patterning of which are shared by two unrelated parrot species that live in tropical savannahs in northern Australia and Africa. He wonders whether there is something about the intense light of tropical savannahs that makes these colours and their patterns advantageous for the birds.

The Blue Gum forests of Tasmania are home to the Swift Parrot (*Lathamus discolor*), a bird that conforms to the green and red norms of parrots but adds splashes of yellow and blue. The Swift Parrot is critically endangered and unusual for being a migratory species. It nests in Tasmania over summer and journeys to live on mainland south-eastern Australia during the non-breeding season.

Swift Parrots share the forests where they breed with sugar gliders, small possums that can jump between distant trees with the help of a fur-covered gliding membrane that stretches between their limbs, like a wingsuit (Plate 17). The sad twist in this story is that they eat Swift Parrots – eggs, chicks and adults.

Dr Clare Holleley leads the vertebrate collections at CSIRO.

'Gliders are omnivores, feeding on eucalyptus gum, sap, insects, nectar, honeydew and lizards,' Clare says. 'It's unusual for birds to make up a large portion of their diet, so we began to wonder, are gliders supposed to be in Tasmania?'

Clare led a team of researchers to find out. They looked into historical records, including early European expedition logs, specimens in Australia's natural history collections and community sightings recorded on the Atlas of Living Australia. Next, they used DNA sequencing to compare Tasmanian gliders with populations from southern Australia.

'The historical data and DNA analysis showed gliders have only been in Tasmania for around 150 years,' Clare says. 'They were introduced from mainland Australia.'

The answer has helped inform conservation managers, allowing them to make evidence-based decisions about how to protect Swift Parrots in Tasmania.

Diving with handfish

A little further south, dwelling at the bottom of the Derwent River estuary in Hobart, is another critically endangered Tasmanian species, the Spotted Handfish (*Brachionichthys hirsutus*).

7 – Space invaders

It's a small, pinkish fish with small brown spots and distinctive fins that are reminiscent of, you guessed it, hands. One of its threats is the Northern Pacific Seastar (*Asterias amurensis*), a pest species that was introduced to the Derwent in the late 1980s in ballast water from a ship.

Helen O'Neill, from the Australian National Fish Collection, works as a diver on the Handfish Conservation Project to help conserve the Spotted Handfish. She tells me the invasive seastars feed on a species called the Stalked Ascidian (*Sycozoa pulchra*), which the handfish use to anchor their eggs. Stalked Ascidian have an otherworldly, plant-like appearance but are actually colonies of animals.

'I call them space tulips,' Helen says.

With the 'space tulips' disappearing, the team plants artificial spawning habitats (ASH). They mimic the ascidians and are made from ceramic.

'There are nine sites where long-term monitoring of Spotted Handfish populations has been conducted. We survey the number of fish and the number of ascidians, and plant ASH where ascidian numbers are low,' Helen says.

Unlike most fish, Spotted Handfish guard their eggs. They hang around the ASH and ward off predators until the eggs hatch.

'From laying to hatching takes 6–10 weeks,' Helen says. 'So we survey the ASH every six weeks to make sure we don't miss a breeding event.'

Helen describes diving in the Derwent River as 'cold and a bit of a moonscape'.

'The handfish live on the muddy, soft, silty bottom. There is a lot of sediment, so we have to dive very carefully so as not to disturb it. If you look closely at the sea floor there are crabs, worms, snails and occasionally you see flathead, skate or stingaree.'

I ask her whether she gets to dive for work in beautiful places, on reefs and tropical islands.

'Not really,' she says. 'Diving sounds glamorous and while it is fun, it's often cold and uncomfortable, particularly during winter, when we do many of our surveys, the water can be 9 degrees.'

Dr Sharon Appleyard is director of the fish collection. She works on genetic studies of the Spotted Handfish. These studies rely on fin clips, which are tiny tissue samples that divers can take from live fish without harming them.

'We used 170 fin clips to look at the population genomics of Spotted Handfish, sourced from seven different sites in the Derwent River, and collected between 2006 and 2008,' Sharon says. 'The fin clips and their DNA are stored in our genetics laboratories.'

The study Sharon led, which was published in 2021, is the only population genomics study undertaken on Spotted Handfish. It supported previous conclusions that individuals tend not to move between locations.

'Eggs are attached to ascidians or ASH, so aren't dispersed by the water currents,' Sharon says. 'Juveniles stay close to where they were born and adults are poor swimmers and don't "walk" far. The genetic results showed the populations cluster into three groups. This suggests the different populations each require conservation to preserve all of the genetic diversity of this very rare species.'

Weeds, bugs and keys

One sunny day I take a short trip with Brendan Lepschi, head curator at the Australian National Herbarium, to collect weeds to use as display specimens. I'll later take them to events and into classrooms, sparing our real specimens from damage while giving kids the opportunity to do hands-on science. Brendan and I pull the car over and jump out to pick specimens of *Tragopogon dubius*, a weedy daisy. Police stop to enquire what we are up to, but they let us continue; there are plenty of weedy daisies to go around.

The herbarium is just as interested in weeds as in native plants because of the biosecurity goal of managing the weeds already present in Australia and keeping others out. Brendan tells me about some of the weedy specimens in the collection.

'*Genista monspessulana*, or Cape Broom, is an established weed in Australia and we have good holdings,' he says. 'It's obvious when it's in flower and it's moderately pretty as a specimen, which means we have multiple specimens. Some are quite old, which can be useful in terms of tracking introduction and spread and helping to understand possible future range expansion.'

Brendan goes on to explain why the herbarium also holds several weedy species that only occur overseas, not in Australia.

'Our collection is worldwide, with representatives of weeds that have so far not been recorded in Australia,' he says. 'In the event we suspect they have arrived here, we have reference material to compare them with.'

Another story he likes to share with visitors to the herbarium is about *Solanum*, the genus that includes tomatoes, potatoes and eggplants.

'*Solanum elaeagnifolium* is a weed that looks superficially similar to native, spiny solanums. It's important to know your native species as well as your weeds. It demonstrates how important it is to know what you are looking at before you start to eradicate it.'

Conversely, it's important to know what you're looking at before you start to conserve it. I'm reminded of this when I run into Dr Alexander Schmidt-Lebuhn, a daisy botanist, in a hallway of the herbarium. Alexander has just 'keyed out' a threatened daisy species and had quite a surprise. It's a weed.

Making identification keys has a long history in taxonomy. As a simple example, imagine you want to use a key to identify an animal. It's small and spikey. A key provides a series of Yes/No questions based on diagnostic features that lead to species identifications. You start at the top and work your way down, like a flowchart. Does it have vertebrae? Yes. Does it have hair? Yes. Does it lay eggs? Yes. Does it have spines? Yes: it's an echidna.

Using a key is more complex when it comes to questions about detailed features like the warts, wrinkles, ridges and reticulations of different daisy seeds. When I first met Alexander, he was making a key to the seeds of weedy daisies.

When potential weed seeds are found at Australia's ports, they need to be identified accurately. The answer reveals whether they are a threatening weed, a weed that has already invaded and become established, or a native species.

'Around 40 per cent of seeds that are found during quarantine inspections belong to daisies,' Alexander says. 'Before our daisy weed seed key, around one in five of couldn't be accurately identified.'

The daisy family, Asteraceae, has more than 25,000 different species. Some we consume, such as lettuce, chamomile and sunflowers. Some are used in the floristry industry, such as gerberas and paper daisies. Many are aggressive weeds. This is in part thanks to their seeds.

'Daisy seeds are often numerous and good travellers because of their parachuting hairs,' Alexander says. 'These are familiar to us from dandelions and formally called pappus. In different species they can be feathery, minutely toothed, smooth, absent or transformed into spines or scales.'

To build the weed seed key, Alexander took seeds from the flower heads of expertly identified herbarium specimens, examined their seeds under a microscope and recorded their features. He travelled to other herbaria in Australia and overseas to access specimens of every species he needed for the key. He examined the seeds under the microscope and recorded their features.

The difference between Alexander's weed seed key and the echidna example above was that in the weed seed key you didn't have to enter at the top and work your way down. You could begin the key with any trait, even using a broken piece of seed missing some of its features.

The key was a success, but was there a faster way to get a result? Something that wouldn't require the user to have any botanical expertise? Alexander photographed the seeds from multiple angles, trained an AI model to recognise them and worked with an app developer. The result was an app that meant anyone could simply hover a smartphone camera over a weed seed to identify it. Then he was asked to do the same thing for the Brown Marmorated Stink Bug (BMSB: *Halyomorpha halys*).

The BMSB is a sap-sucking bug native to eastern Asia. It is a threat to crops such as apples, stone fruits, hazelnuts and grains. The bugs are attracted to light, but pregnant females like to hide in dark spaces. Shipments of pottery or new cars are perfect for places for pregnant females to stow away.

Like many bugs and beetles on the biosecurity risk list, the BMSB has many native Australian lookalikes. Australia has around 600 scientifically named species of stink bugs and several thousand more awaiting classification. Alexander used stink bugs in the insect collection to train an AI model to recognise the BMSB and tell it apart from harmless natives. The AI model was then included in an app.

'The app uses the video feed of a smartphone,' Alexander says. 'You hold the camera over the bug and can zoom in or out and look at it from different angles. The AI model in the app provides an identification estimate and its degree of certainty.'

'The important thing is to be able to tell the difference between the BMSB and other bugs. No one wants to delay or reject a shipment over a native bug.'

An endangered weed

Mistaken identity is not uncommon in the world of weedy daisies. Fireweed (*Senecio madagascariensis*) is one example. It's native to southern Africa and highly invasive in many parts of the world. In Australia it's a Weed of National Significance, one of 32 weeds recognised for their invasiveness and harmful environmental, social and economic impacts. But Fireweed wasn't always recognised as a weed.

'This plant was considered native in Australia until the 1980s,' Alexander says. 'People thought it was part of the very variable *Senecio pinnatifolius* complex, until somebody got suspicious and consulted with a South African taxonomist.

Then it remained controversial until somebody did DNA sequencing and chromosome counts and confirmed it in the late 1990s.'

Fireweed wasn't thought to be present in New Zealand until very recently. Some of its relatives are also weeds, including the Gravel Groundsel (*S. skirrhodon*). It's much less of a problem than Fireweed, but the two species are difficult to tell apart.

'Their flowers look similar,' Alexander says. 'And leaf shapes vary a lot, even within the same species, so it is difficult to identify them.'

Working with a collaborator in New Zealand, who suspected Fireweed had gone undetected there, Alexander and colleagues used DNA sequencing to compare weed samples from New Zealand with Fireweed specimens held in herbaria. They used the DNA sequences to build a phylogenetic tree, showing the relationships of the plants.

'It turned out that some of New Zealand's supposed Gravel Groundsel was actually Fireweed,' Alexander says.

The phylogenetic tree also hinted at the history of the Fireweed invasion of New Zealand, showing it may have spread there from Australia.

On the day I run into a surprised Alexander in a hallway of the herbarium, he's just keyed out a daisy specimen and got an unexpected result. It's the very rare Daisy Fleabane (*Erigeron conyzoides*). It's on Victoria's threatened species list, but Alexander has just keyed it out as a weed.

Alexander and postdoctoral researcher Dr Stephanie Chen are working on the taxonomy of Australian relatives of the invasive weed Flaxleaf Fleabane (*E. bonariensis*). This kind of work is important for ensuring that biocontrol agents such as fungi and insects will attack only the target weed, not local species, before they are released into Australia. A first step is getting a grip on what the local species actually are.

A few weeks later, Alexander sends me a photo of Daisy Fleabane growing on a dam wall (Plate 18). Alexander and Stephanie have either just recorded the first sighting of this rare species in 13 years, or they've found a weed. DNA will tell.

'It will be interesting if our genetic data show it to be introduced,' Alexander says. 'Either we found a very rare plant, or a weed has been mistaken for a native since 1854.'

When I next catch up with him, he has the answer. Daisy Fleabane is a weed.

'Both DNA data and morphology indicate that the rare plant *E. conyzoides* is actually *E. acer*, a species from the northern hemisphere,' Alexander says.

He speculates that it arrived in Australia with migrants from Sweden or Switzerland and found new places to grow in alpine areas of New South Wales and Victoria. We look at the photo of it growing in a weed community on a dam wall.

'There were no plants of this species growing in the native vegetation nearby,' he says, referring to the weedy daisies' habit of invading disturbed areas, like roadsides.

Alexander notes the irony of the finding. Daisy Fleabane had been considered the most important Australian relative of the invasive weed Flaxleaf Fleabane. Any biocontrol agents released to control it would have needed to be tested to ensure they didn't harm Daisy Fleabane.

There are so many quirky tales across the collections, not to mention quirky tails ...

8

Tales of genitalia ... and other strange ways to get information about animals

'What do you do with your genitalia?' In most contexts, such a question would be inappropriate. But if two entomologists meet for coffee, it's not an extraordinary question.

Quirky tails

Towards the end of writing this book, I put out a call for quirky tales, preferably involving the tail ends of creatures. The first person to reply is Dr Dan Huston, whom we've met twice in this book, once when he named a parasite after his baby and again when he worked on the National Diagnostic Protocol for Pine Wood Nematode.

Dan's not quite sure whether his work fits the brief and has attached an image to his email. I open it to find a black and white lunar landscape showing a crevasse surrounded by ripples. But the scale bar indicates the crevasse is only 100th of a millimetre.

'Not sure if it's quite quirky enough,' Dan writes, 'but I am currently in the process of building a large collection of root-knot nematode perineal patterns.'

Root-knot nematodes live in soil and parasitise the roots of plants, causing problems for agriculture.

'The perineal pattern is a fingerprint-like pattern around a female root-knot nematode's vulva,' Dan continues in his email. 'They are generally thought to be species-specific, so you can use the pattern to identify a species, assuming you can make the right interpretation!'

'The female root-knot nematodes are about half a millimetre in diameter. I have to cut the perineum off the back end of the nematode and mount it flat on a microscope slide. In my current project I'd like to use AI to identify these nematodes to species, because it is hard to do as a human!'

In addition to Dan's root-knot nematode vulvas, the Australian National Insect Collection holds more than 50,000 microscope slides. Some are whole creatures, such as tiny flies. Others are genitalia, many of which belong to moths and butterflies. The paper describing the Sapphire Azure butterfly that we met in Chapter 2 features beautiful photos of the bright blue wings and scales of species in the genus. It also has photos of detailed dissections of male genitalia, allowing careful comparisons between different species.

Dr Federica Turco manages the insect collection, a role that involves overseeing the care and curation of more than 12 million specimens. Earlier in her career she specialised in Meloidae, a family also known as blister beetles.

'Blister beetles feature rather boring male genitalia all over the world,' Federica says. 'The exception is Australia. Here they have gone crazy with hooks and anchor-shaped parts.'

Federica shows me some of her scientific drawings. They show the male genitalia of blister beetle species from the Northern Territory and Western Australia, which have complex-looking hooks on the ends. She explains they probably help hold the males and females together during mating, but that is yet to be determined.

'During mating, many insects have what's called a linear phase,' Federica says. 'You might sometimes notice a pair of insects joined together end to end; they can stay like that for hours. But this is not usually the case for beetles, which mate quickly with the male on the back of the female.'

'Blister beetles found elsewhere in the world, featuring less flamboyant structures, are exceptions to this rule and also show end-to-end mating. The incredibly complex hooks found in the Australian meloids leads the imagination to wonder. That is, until scientific evidence emerges.'

Federica tells me that sexual behaviour and morphology (the physical form) of genitalia can be critical to speciation, the process that leads to the evolution of new species. This is why taxonomists are so interested in genitalia.

In an interesting aside, she mentions that blister beetles parasitise solitary bee species. The tiny larva of a blister beetle is very mobile. It hatches from an egg laid underground, climbs onto a flower, uses its claws to grab onto a visiting bee then lives in the bee's nest, eating its larvae and food provisions.

Blister beetles often have bright metallic colours, a warning to predators to stay away. They produce a chemical called cantharidin in their haemolymph (insect blood) and release it from their joints in droplets when stressed. Cantharidin can stain skin and cause blisters on mucosal surfaces like the mouth and eyes – hence the beetles' name.

More tales of tail ends

So crucial to taxonomy are the tail ends of insects that they were the key to ridding the country of a particularly noxious weed. *Salvinia molesta* is a floating fern that chokes lakes and rivers. By the late 1970s it was such a problem that CSIRO began to search for natural enemies in its original home in Brazil. Careful observation led to scientific interest in what was then thought to be one species of weevil, *Cyrtobagous singularis*, that lived on the plant. Live specimens were exported to Brisbane, where ecologist Dr Don Sands tested them on *Salvinia* plants growing in the laboratory. Some of the weevil larvae simply crawled around on the plants but others tunnelled into the stems, killing them. The latter were bred, cultured, safety tested with non-target plants and then released. This led to a dramatic reduction in *Salvinia*, a success that was later rolled out in other countries and resulted in a colour picture on the cover of the scientific journal *Nature*.

Meanwhile, Don looked more closely at the two kinds of weevils, those that crawled on the plants underwater versus those that tunnelled into the stems underwater. The crawlers had long hairs that helped them cling to hair-like structures on the plants and not fall off. The tunnellers didn't need to hold on, and had short hairs. The difference was obvious in their genitalia – they were two different species. Don and a colleague described and named the tunnelling species *C. salviniae* in 1986.

Next, Dr Michael Frese phones to trump the caddisfly, sawfly and jumping spider fossils we met earlier in this book. The fossil site at McGraths Flat has preserved a nematode along with a mite. The mite is only ~1 mm long and sits right on top of the nematode.

'Perhaps, the mite was chasing the nematode when both found their untimely death in the oxbow lake at the centre of the fossil deposit,' Michael says.

I'm typing this section when one of my teenage sons suddenly asks whether I'm writing about worm bums again. It reminds me to include the strangest-looking worm of all, a peanut worm collected during a voyage on board RV *Investigator* that bore a close resemblance to a part of the human male anatomy. Peanut worms are marine creatures in a taxonomic group called Annelida, the segmented worms, which also includes earthworms and leeches. The nematodes we've met so far in this book belong to a different group, Nematoda, the roundworms.

As one story always leads to another, let's talk about another annelid. Working with the insect collection, in 2021 Dr Geoff Dyne described a giant earthworm in a new genus, *Aridulodrilus molesworthae*. It's 1.5 m long when extended. Not only is it unusually long for an earthworm, it lives in a semi-arid landscape near Broken Hill, NSW, not in the moist soils in which we normally find earthworms. Australia has around 750 named species of earthworms and their ancestors probably first appeared on Earth around 200 million years ago.

The love lives of sharks

About an hour after Dan has introduced me to the perineal patterns of tiny nematodes, Helen O'Neill from the Australian National Fish Collection sends me something from the opposite end of the scale. It's a photo of herself standing next to a structure on a laboratory bench. The strange-looking structure is at least half her height and resembles the weathered remains of a gum tree trunk. It's the penis of a Basking Shark (*Cetorhinus maximus*).

We arrange to meet online. During our meeting Helen shows me some strange photos of shark genitalia and sharks

Wild Collections

postured in unusual positions. We're pretty sure our activity will be flagged by our IT people and we'll have some explaining to do.

The penises of sharks and rays are called claspers. And they have two of them.

'A clasper is a modified portion of the pelvic fin,' Helen says. 'There are two pelvic fins, so two claspers. They can be unfurled, sort of like an umbrella.'

Helen shows me another photo from the fish collection of a clasper being unrolled and pinned in place (Plate 19). It does indeed look like a rolled-up fin. During mating, the male shark or ray inserts one of its claspers into the female's cloaca and a tube running through the clasper delivers sperm.

'In juvenile male sharks, the claspers are soft,' Helen says. 'In adolescents they are semi-hard and in adults they are hard due to them becoming calcified as they mature. The photo I sent you of the Basking Shark's claspers are the largest in the collection. They do look quite weathered.'

'Male sharks bite the females while mating to help them stay in place. It must be hard to have sex under water without hands to hold onto things,' Helen says. 'I sometimes see female sharks in the collection that have marks and scars from mating. Females have actually evolved thicker skin than males to cope with bites during mating.'

Helen mentions that male chimaeras, which are related to sharks and rays, have an extra clasper on their heads to help them hold on to females.

Does this mean they have you-know-whats on their heads? We look at some photos. The head clasper, or rather the frontal tentaculum, is a small protuberance with little angled spikes on the underside. Helen has handled some male chimaeras in the fish collection and confirms that yes, those spikes are quite sharp and can hook onto your gloves. These are used to hold onto the female's pectoral fin to help keep them in place while mating.

Unlike sharks and their relatives, most bony fish in the sea practise external fertilisation. The females lay eggs, the males add sperm.

The exception to this rule is blind cusk eels, a family that practises internal fertilisation. Helen shows me some photos of specimens that were collected during the Gascoyne voyage on board RV *Investigator*, the same voyage that discovered cat sharks living in sponge hotels (Chapter 2). The photos of the blind cusk eels show small fish around 7 cm long. Male blind cusk eels have a fleshy penis that is usually externally visible. Clusters of eggs in the abdomens of females can sometimes be seen through transparent skin.

My conversation with Helen makes it sound like sexing a shark should be easy – simply look for the claspers. But this can be difficult in shark embryos and impossible if all you have is a tissue sample or a fin from a fish market.

Dr Floriaan Devloo-Delva is a research scientist at the fish collection. He has developed a PCR test to determine the sex of a White Shark (*Carcharodon carcharias*) that requires only a DNA sample extracted from a tiny piece of tissue. But first, Floriaan needed to find out how sex is determined in this species.

Sex determination varies widely in the animal kingdom. It is not always driven by sex chromosomes. Sometimes egg incubation temperature is what determines the sex of the offspring, such as in crocodiles, or it can override sex chromosomes, such as in some species of lizards. The molecular mechanisms that determine sex in sharks are mostly unknown.

Floriaan and the team he worked with developed a statistical method to search a shark's DNA for sex chromosomes.

'We discovered that sex determination in the White Shark is driven by genetics, not by temperature like in crocodiles or turtles,' Floriaan says. 'The White Shark has X and Y sex chromosomes. Males are XY and females are XX.'

White Sharks are feared by many beach goers, but are listed as Vulnerable to extinction on the IUCN Red List. Understanding more about their biology and being able to determine the proportion of males and females in a population is important for managing the species.

'Our new test will help population monitoring of the White Shark by genetically identifying the sex when the claspers were not easily visible during biopsy sampling from the boat or dive cage,' Floriaan says. 'It can also be used on samples obtained from processed carcasses in fisheries or the fin trade.'

Wild tales of reproduction

Next to reply to my email is Ashara Patterson, who is digitising orchids at the Australian National Herbarium. In a previous job she bred insects in a quarantine facility for use in biocontrol research. But the insects weren't always willing.

'We had a couple of weevil species and a moth called *Dichrorampha aeratana*,' Ashara writes. 'Sometimes they just didn't want to continue the population. Some days if I came into work and found a pair mating, I would excitedly tell the whole team that they were making babies. They would be almost as thrilled as me!'

Ashara has attached photos of her weevil babies to her email. They are as cute as buttons.

Also on the topic of reproduction is a quirky tale from Dr Laurence Mound, an honorary fellow from the insect collection who studies thrips. These tiny insects make up an order of insects called Thysanoptera and most are less than 1 mm long. Laurence suggests thrips associated with Australian *Acacia* trees might not sound exciting, but the way they reproduce is.

'In these thrips, the first eggs laid in a gall develop into wingless soldier adults. These protect their mother's developing second generation in the gall, permitting them to grow up as normal winged adults. But the soldier generation, in accepting the responsibility of protecting their younger siblings, has altruistically forgone their own right to reproduce.'

'The funding objective was to explore the forces driving this rare self-sacrificial behaviour. But it also provided the opportunity to explore the Australian thrips diversity on *Acacia*, resulting in over 200 new species and 20 new genera.'

Not to be outdone, the Australian Tree Seed Centre joins in. Tasha James has worked across the collections, first as a herbarium trainee, then as a technician supporting the big move (see Chapter 9) and now with tree seeds. Like the Australian National Algae Culture Collection, the tree seeds are alive. But growing a tree is not as simple a placing a seed in soil. Different tree seeds require different pre-treatments.

'For the propagation of many *Acacia* species, we boil the seeds to trigger their germination,' Tasha says. 'This is affectionately known as making "Acacia tea". We even use tea strainers to hold the seeds.'

'We test seed regularly using a variety of seed germination techniques tailored to each species.'

Tasha shows me a photo of a mini forest in her laboratory. It looks like a terrarium display, but it's actually a seed viability test.

'Seed leads to living giants,' Tasha says. 'Many of our species are large forest trees that begin as one tiny seed in a handful of thousands. Trees such as these are not only incredible lifeforms themselves, but also provide habitats for other plants and animals.'

Australia supplies seed to countries across Europe, Africa, the Pacific and Asia for purposes such as conservation, research, forestry, seed orchards, food, nurseries and many more.

'We also work with people in developing countries to produce timber for many uses, including building material for income and houses, and biomass.'

Buckets of gunk

The next email I receive is from Dr Erin Hahn, a conservation geneticist who works with the Australian National Wildlife Collection. She sends me a photo of herself wearing a long yellow glove on her arm that is disappearing into a large white bucket. I assume it contains reptiles stored in ethanol.

Erin and Dr Clare Holleley, who leads the vertebrate collections, have discovered a way to get a unique kind of information out of old museum specimens that were preserved in formalin. This preservative locks up DNA, a fact that had made the millions of reptile, amphibian and fish specimens stored in the world's collections unusable for DNA sequencing. But Clare reasoned that formalin is a first step in many laboratory experiments involving DNA. What if she looked at specimens preserved in formalin as if they were really just DNA experiments that started 10 years ago, or even 100 years ago? It turns out she was onto something.

When genes are switched on inside a cell, their normally tightly wound DNA opens to allow cell machinery to read the DNA's code. Formalin preserves this open position by crosslinking DNA to the proteins associated with it. Clare and Erin developed a way to read this information. It reveals which genes were switched on at the time the animal died, showing how that animal was responding to its environment at the time.

'We can read this information even in very old specimens,' Erin says. 'The oldest so far was an Eastern Water Dragon collected in Queensland in 1905.'

'The next step is to compare specimens collected across time in different places and match this to environmental information. It would show how animals have responded to environmental change in the past. We're interested in this because of the rapid environmental change animals face now.'

Formalin has suddenly gone from being a problem for DNA research, to an asset. And it doesn't stop with animal specimens.

'Tissue banks around the world hold medical specimens preserved using formalin,' Erin says. 'Our technique could show the influence of environmental factors on diseases like cancer.'

Erin is also using wildlife specimens stored in ethanol for a different type of DNA study, or rather she is using the ethanol itself. Many of the specimens in ethanol in the wildlife collection are stored together, not individually, in large jars or buckets. From time to time, the ethanol is replaced. It would normally be disposed of as chemical waste, but Erin realised it might hold DNA of the specimens stored in it, or even the DNA of the parasites and microbes that might be associated with them. Erin thinks working with buckets of gunk that other people would throw away is quirky enough for this chapter and I agree.

Speaking of buckets, I join PhD student Kate O'Hara and postdoctoral researcher Dr Livia Gerber at the wildlife collection to film Kate teaching Livia how to dissect livers from lizard specimens. They begin by carrying a bucket of the large lizards Livia will be working on down the hallway from the ethanol vault to a laboratory with a fume hood.

Kate is an evolutionary biologist who works on Bynoe's Gecko (*Heteronotia binoei*). Some populations of this species are all female. They reproduce by parthenogenesis, which basically means they clone themselves. These all-female populations are like identical twins, but there are thousands of them.

'The big question is how does this species cope with environmental change?' Kate says. 'The all-female populations have very little genetic diversity.'

Diversity is thought to be a major driver behind sexual reproduction. If, for example, a gecko produces offspring that are different from each other, chances are some of them will survive a virus, live through a drought or outrun a predator thanks to those differences. If their offspring are identical, what's the impact on survival?

Kate's looking for answers in the DNA of specimens of *H. binoei* held in the wildlife collection. I film her carefully dissecting liver samples from specimens stored in ethanol. She places them in individual tubes, ready for DNA extraction. The geckoes are only around 10 cm long, including their tails. They were preserved in formalin before being stored in ethanol, so Kate will build on Erin's and Clare's work to dive into the stories in their DNA.

Similarly, Livia is working with much larger specimens of Central Bearded Dragons (*Pogona vitticeps*), dissecting their livers to extract DNA. She is applying the same techniques to investigate how temperature interacts with bearded dragons' genomes to impact their lifespan.

'My research helps understand how cold-blooded species respond to climate change,' Livia says. 'This will help people develop better wildlife conservation strategies while also teaching us about aging, a process that we all go through whether we like it or not.'

As strange as all this may seem, Erin and others have far stranger ways of getting their hands on DNA.

Out of thin air

Wherever we go, we leave DNA. Forensic traces of the code that defines us linger like footprints. We even shed our DNA into the air.

Plants, animals and other microorganisms shed DNA into the environment too. Collected in samples of water, soil or air, this environmental DNA, or eDNA, can be sequenced to identify the species present in the area. It's a non-invasive way to study animals without disturbing them, or even seeing them.

Airborne eDNA is the newest kid on the block. The rigs to collect it can be super high tech with in-built weather monitoring capabilities or as simple as a piece of poly pipe with a piece of filter paper inside, hung from a tree by a piece of string. Airborne eDNA monitoring could give us snapshots of biodiversity or help detect the entry of invasive species, such as rabbits and foxes, into a wildlife reserve.

'We're trialling the concept at a zoo,' Erin says. 'We're attempting to detect the DNA of the zoo's inhabitants to better understand how DNA moves through the air so we can track all species from Shingleback Lizards to rhinos.'

Detecting eDNA in sea water is a much more mature technology that is used to measure and monitor fish biodiversity. In 2023 Dr Haylea Power, a research scientist at CSIRO, decided to find out exactly what people were collecting when they scooped up sea water samples to collect eDNA. Were there free-floating pieces of DNA in the water or was the situation a bit more complex?

'We know how to collect eDNA and work with it in the laboratory, but we never knew exactly what we were working with,' Haylea says.

Haylea's team began by collecting sea water samples from Mettam's Pool, a lagoon in Perth, Western Australia. That involved simply snorkelling in the water and filling large plastic jars with sea water. Back in the laboratory, they filtered the sea water through a series of filter papers with progressively smaller pores. This separated particles of different sizes onto the different filter papers. They used microscopy to visualise those particles then analysed the eDNA attached to them.

They found naked eDNA can be present in sea water, but that isn't the whole story.

'eDNA is present in a variety of different ways in sea water,' Haylea says. 'It can be free-floating fragments of DNA, DNA inside single cells or tissue fragments, whole microscopic organisms or a mix of these things embedded in biofilms.'

Dr Cindy Bessey, a research scientist at CSIRO who uses eDNA for marine biodiversity studies, had been working with eDNA for a long time when she began to wonder whether there was an alternative to spending many hours at sea in a small dinghy with a water filter strapped to her back, filtering thousands of litres of sea water. There was. Simply floating filter paper in the sea gets the same results.

'We call this passive sampling,' Cindy says. 'It's no longer necessary to filter sea water to collect eDNA. As well as saving time, it means eDNA studies can be done in places where access to equipment or power is limited.'

Cindy is now working on an eDNA collector that attaches to RV *Investigator*'s deep towed camera, a system designed to provide imagery of deep sea features and biodiversity. The eDNA collector goes down closed, opens at depth and collects eDNA by water flowing over filter papers. The next step is to compare the species detected by watching the camera footage with the species detected by analysing the eDNA.

Dr Katrina West, a geneticist at the fish collection, tells me about her recent study where the team compared eDNA sampling with observations by divers on shallow Tasmanian reefs. The field trips involved collecting data for both methods over a 24 hour period, which meant putting on wetsuits and getting in the water at 10 a.m., 4 p.m., 10 p.m. and 4 a.m.

'Different fish come out at night, which means the timing of a dive survey matters,' Katrina says. 'The signal from eDNA is more consistent. Fish only seen at night are still detected in daytime eDNA and vice versa.'

Other staff at the fish collection have supplied tissue samples from the expertly identified specimens in the collection to CSIRO's project to build a National Biodiversity DNA Library (NBDL). eDNA is most useful if it can be matched to the species that shed it, which is possible only if there's a reference DNA sequence for comparison. The NBDL will be a reference for all scientifically named Australian species of animals, plants and some microbes.

Cindy and Haylea sometimes swap the deep ocean for the Aquarium of Western Australia's 3 million litre tank, the largest single aquarium in Australia. It simulates the natural environment, but unlike the natural environment there's a list of its inhabitants. Cindy has tested different materials for collecting eDNA, such as hemp and synthetic sponge, and Haylea is developing ways to count fish by using whole cells shed into water. One day it will be possible to find out from just a bucket of sea water not only which species are present in the area, but also how many of each.

Perhaps the strangest example comes from Dr Nerida Wilson, who works with the NBDL. Her expertise is marine invertebrates and in a previous job she used an unusual technique to sample DNA from a Dana Octopus Squid, the species we met in Chapter 5. Nerida shows me a video of the squid, filmed underwater during an expedition she led to Ningaloo Reef in 2020. The squid is interacting with a strange-looking brush on the end of a robotic arm.

'We managed to get DNA from it by getting it to engage with a kitchen brush and taking samples from the stiff bristles,' Nerida says. 'The online community that was watching us live streaming that called it the KBOS, Kitchen Brush of Science, and it stuck.'

Dr Alyssa Budd is another postdoctoral researcher at CSIRO who is using DNA as a source of information for wildlife management, but in a slightly different way. She has built on earlier research that used genome sequences to estimate the natural lifespan of vertebrates. It's based on the density of certain DNA sequences called CpG sites.

Alyssa's method estimates the age of sexual maturity for any vertebrate species, using only its genome sequence. It's a critical factor for managing wild populations, such as fish stocks. Fish need to survive long enough to reproduce or the fishery won't be sustainable.

'Getting age at maturity data has always been difficult,' Alyssa says. 'You can observe animal populations in the wild over a long period, waiting to see when they start to produce offspring, or you can analyse the gonads under a microscope to find out if they contain viable egg or sperm cells. Even then, you still need to know their age, which is another big challenge.'

'Or you can take a tiny tissue sample like a fin clip from just one individual, sequence the genome, and use our model to predict the age at maturity for the entire species.'

Both methods – lifespan and age at maturity – can give estimates about extinct species, sometimes by extrapolating from the genomes of living relatives. Mammoths likely lived for around 60 years. Neanderthals and early modern humans lived for around 38 years.

Jurassic lark

My request for quirky tales yields one final offer. Dr David Yeates wonders if I'd like to know whether mosquitoes bit dinosaurs. Of course I want to know this – everyone does! And we have a follow-up question: can you sequence the DNA of dinosaurs?

I meet with David in his office and he hands me a paper titled 'Phylogenomics reveals the history of host use in mosquitoes'. It was published in *Nature Communications* in 2023 and David is one of its more than 20 authors.

The purpose of this research was not to establish whether mosquitoes drank the blood of dinosaurs. It was to use DNA sequencing to resolve the phylogeny of mosquitoes, meaning figuring out the relationships between them. The work is especially important to further our understanding of the more than 100 mosquito species known to spread diseases to humans, and the further 200 or so that might.

The analysis showed mosquitoes are much older than previously thought. The earliest mosquitoes likely lived in the early Triassic, around 217 million years ago. Analysis of the different lineages and the hosts they feed on today suggest these ancient ancestors probably drank the blood of amphibians. Blood meals are necessary for female mosquitoes to produce eggs. The rest of the time, mosquitoes feed on nectar, rotting fruit and so on.

'Mosquitoes are old enough to have fed on dinosaurs,' David says. 'They feed on all their living relatives today, including crocodiles and birds, so we can assume they also fed on dinosaurs. Birds are the living descendants of one clade of dinosaurs. Crocodiles are tough, but mosquitoes target the chinks in their armour, such as around their eyes.'

Can we get DNA out of mosquitoes fossilised in amber more than 66 million years ago and use it to resurrect a dinosaur? David says even the first step is unlikely.

'There seems to be no DNA, even in younger amber,' he says. 'People have tried extracting DNA from big termite inclusions in Baltic Amber, around 25 million years old, without success. We'll probably never be able to sequence any DNA more than a few million years old.'

I notice a photo on David's desk. A large moth appears to have a dozen or so long twigs growing out of its body. But they aren't twigs. The moth has been attacked by a cordyceps fungus and no one at the insect collection has seen anything exactly like it before. The member of the public who found the deceased moth has it waiting in their freezer, in Queensland. David's going to send someone there to carry it carefully back to the insect collection, but it's been hard to find the time.

The insect collection is in the middle of moving house and there are 12 million residents to relocate.

9
Big moves and quantum leaps

For most people, the thought of moving house is a little overwhelming. Now imagine you have more than 13 million individual items, many as fragile as a butterfly's wing and as alluring to pest insects as hot chips to a flock of seagulls.

A new home

It's the end of July 2024, a typically sunny morning in Canberra, but so unseasonably warm that I've left my long puffer jacket at home for the first time this winter. I join a stream of people walking across the CSIRO site towards our new collections building, named Diversity (Plate 20).

When I reach the foyer, the mood is jubilant. We're finally standing in the building we'd only imagined through architects' drawings or experienced the handful of times we visited during construction. It's beautiful. Earthy tones, vast expanses of glass, and views from the laboratories and meeting rooms. Who wouldn't want to work here? Why would you ever go home?

While we put blue cheesecloth booties over our shoes to keep the building clean, a few of us try to remember when we began to plan this building. How long ago was it that we all joined the affectionately named FUGs (functional user groups) so the architects could co-design the building with the people who would use it? There were FUGs for cryogenics, loans, collection management and more. My FUG for outreach and display lost out to the reality that opening the vaults to the public and TV crews would risk also opening them to pest insects.

While we're gathered in the foyer, our director gives an Acknowledgment of Ngunnawal Country, followed by a safety briefing. Then we're free to explore the building.

I decide to make myself familiar with the route of the staff tour I'm leading later that day. This turned out to be a very sensible precaution. I'm having trouble correlating my mental map of the layout with the reality of the completed building. It feels to me as if the tearoom is on the wrong floor and the laboratories and vaults are reversed. I've no idea how to get to the receivals hall, where new specimens will arrive freshly collected from the field and loaned specimens will return from collections in Australia and around the world.

I climb the stairs to the third floor and turn left into the Australian National Insect Collection's wing. A vast corridor leads to the far wall of the building. To the left are curation hubs, with a wall of windows and benches for computers, microscopes and specimens. To the right is the enormous collection vault, accessed through two anterooms that function like an air lock. A group of us enters and waits for the door to the corridor to close, but its sensor keeps reacting to our movements and the door won't close. We whisper to each other to freeze. The door locks with a click and someone opens the vault.

It's whiter than white inside: white floor, white walls, white ceiling and shiny white cabinets to hold the collection. A white fabric airduct around the perimeter of the ceiling creates positive air pressure to help keep out pest insects. I slip my sunglasses down from the top of my head as someone has a crack about the white room in the 1971 film *Willy Wonka & the Chocolate Factory*.

The space is vast. I imagine we could park a jet in here, but my perception of scale is not to be trusted. This vault will hold the majority of the insect collection, about 11 million specimens, at a cool 16°C, thanks to the thick concrete walls. In a worst case scenario – a failure of the cooling system in midsummer – the vault would maintain a stable temperature for several days to protect the specimens.

The mechanical plant system that controls the building is isolated from the working spaces, with separate access like the servants' quarters in an English country house. As I continue my trek around the building, I catch glimpses of metal staircases behind doors that are off limits to me.

I head back to the main staircase in the foyer and walk down one floor. To my right is a meeting room set up like a board room. It has a window wall. I make a mental note to hold all my meetings in here.

Immediately below the insect collection is the Australian National Wildlife Collection, almost a replica of the floor above. To the left of the corridor are spaces for curation and office work. To the right are the wildlife collection's vaults. The cabinets are designed to hold a variety of specimens: trays of birds and mammals; bird eggs (Plate 21); boxes of skeletons; and pelts of dingos and kangaroos. There are even a couple of big cats, which were confiscated on arrival in Australia after a failed attempt to smuggle them in.

On the same floor is the digitisation laboratory. It's a huge, bright room that overlooks the garden at the front of the building. There are a dozen or more large workstations where cameras will be set up to take high-resolution photographs of specimens, as part of the push to get the collections online and free to access for research, education and environmental management. This bright, happy laboratory is a huge step up from the cupboard-like spaces people have been working in.

Down another floor are the molecular laboratories, where researchers will work with DNA using technologies like polymerase chain reaction (PCR), which the public became familiar with during the COVID-19 pandemic as they waited for the daily tallies of infections diagnosed using PCR to amplify viral DNA in nasal swabs.

I wonder how I'll find anyone in this building once the specimens and staff have moved in. It's so much bigger than we could have hoped for, which is testament to how valuable CSIRO's collections are for research.

Opposite the molecular laboratory is the Trace laboratory. CSIRO has never had a space like this before. It's for working with tiny amounts of DNA, the sort of traces you might find in the bones of prehistoric creatures or in filters placed to detect invasive mammals from DNA floating in the air, as we learnt in Chapter 8. A researcher might spend all day working in a Trace laboratory, so it's been designed with thick, double-glazed windows so they can glimpse the outside world. There are separate entrances for people and laboratory supplies. People can shower, then change into a mask, safety glasses, gloves, outer suit and boots. Laboratory supplies will enter through a hatch that looks a little like a servery. It has glass doors opening from the corridor to the laboratory and a UV light to destroy any DNA that could contaminate the laboratory.

Next door is the ethanol vault, a room designed with a tank under the floor to collect spills should the building experience some kind of catastrophe that breaks the specimen jars. Later, I will visit this room with Clare Holleley and write about it in the opening chapter of this book. But today I just look through the door at the empty shelves, and dream of holding public tours here. These specimens are less vulnerable to pests than those stored in the insect and wildlife vaults, which will be off limits to everyone except essential staff, research visitors and collaborators. The labels can grow mould, though, and this is something the curators will keep a close eye on. A specimen with no label, or a label eaten by mould, is no use to anyone.

Further along the corridor I discover a large room being vacuumed. Cleaners and tradies are still working around the building, putting on finishing touches and dealing with minor defects the architect has marked with little pieces of blue tape. This room has doors opening to the outside and several walk-through freezers opening to the corridor. I realise it must be the receivals hall.

I track back through the building to where its three levels connect across an internal bridge to the three levels of the existing building of the Australian National Herbarium. The new collections building curves here to accommodate an old oak tree. A few people are gathered at the glass wall, excited to see the oak has produced buds in anticipation of spring. It seems to have survived the construction of the new building. When warmer weather comes we will eat lunch out here and talk about the irony that our beloved oak is an exotic species.

Pinning down the move

Preparing specimens for the move to the new collections building began several years ago. The Australian National

Insect Collection was founded in the late 1920s and today holds more than 12 million specimens. Some are on microscope slides or in vials of ethanol, but most are pinned and kept in drawers. They represent a phenomenal amount of work, something I realise when I attend a gathering of the Moths and Butterflies of Australasia (MABA) organisation.

There's a mix of students, hobbyists and scientists at the MABA's annual weekend. Many of them specialise in other scientific fields and do this just for fun. To them it's a hobby, like playing golf or painting landscapes. But the weekend is for learning, and I discover that there is much more to pinning a moth, or 'setting' a moth, than simply sticking a pin through its thorax.

The weekend begins in bushland on a Friday night, where we set up a few of what must be science's most simple experimental rigs: a white sheet with a light behind it. You can try this at home with a bed sheet and a torch.

Many insects will come to light, like moths to a flame. You've probably heard the myth that moths are attracted to light because they use the moon to navigate. In 2024, the authors of a paper in *Nature Communications* debunked this idea after using high-resolution video to film how insects behave around light. They revealed insects that appear to be attracted to light have a dorsal light response, meaning they orient their backs towards light. If the light comes from the moon, a dorsal light response helps them to fly. If the light comes from a streetlight or the lamp outside your front door, a dorsal light response traps them in an erratic flight path around the artificial light. It turns out moths are not drawn to flames, they're entrapped by flames.

The people I'm spending the weekend with are drawn to light. They gather in front of the white sheets and get excited about the many different species that land on them. I watch as a university student catches a black and orange moth – his favourite colours – by scooping it off the sheet into a 50 mL Falcon tube, a classic piece of laboratory equipment with graduated markings and a blue lid. Many of the moths caught tonight will be let go after a few minutes, but a few will be frozen to keep as specimens for the setting demonstration later in the weekend.

On Sunday afternoon the same group gathers indoors to learn how to set the moths they collected on Friday night. It's an extremely delicate process that is complicated by rigor mortis, which stiffens the muscles that control the moth's wings. A wrong move can cause the wings to snap right off. Moths can sometimes be held in a humid chamber to soften them before they are pinned.

An expert demonstrates how to position a moth on a setting board, place a pin through its thorax, gently tease its wings into the open position and hold them in place with fine strips of plastic that will be removed in a few days when the moth stiffens in the new position.

I walk around the room watching people trying to create a well pinned specimen. People working with small moths use dissecting microscopes, the kind that don't hold microscope slides but instead give a magnified view of a three-dimensional specimen. Experienced setters give advice on things like correct wing angles and how to use fine forceps to fan open the wings. One person demonstrates a much faster technique that involves using a length of sewing cotton to fan the moth's wings open. The others gather around him, transfixed.

I'm beginning to appreciate the century of collecting and curating that has gone into accruing the vast number of Lepidoptera (moth and butterfly) specimens in the insect collection. They are stored like the rest of the insect collection, in foam-lined cardboard boxes arranged in shallow metal drawers with glass lids, which slide into storage cabinets. The drawers are moving to the new collections building, but the cabinets are not. Neither is the naphthalene (the chemical in mothballs) inside the drawers, which for decades has been used to protect the insect collection by deterring pest insects. The new vaults will do the job by isolating the specimens from the world.

To move the drawers to the new collections building, they need to be packed, transported, frozen on site to kill pest insects in all their stages of life, and placed inside the new cabinets in the vaults. It's not a simple process.

Every time I drop into the insect collection, I notice people preparing for the move. One day a person is vacuuming naphthalene out of drawers, working inside a fume hood to avoid breathing it in. Another day a person is tidying up tiny pins. It's Thekla Pleines, a curatorial technician who has been transferring moths and butterflies out of old-style wooden storage boxes. It's a job she's spent years working on.

'When I joined the Lepidoptera collection in 2018, specimen transfer became one of my main duties, especially in the first year,' Thekla says. 'I owe a huge thank you to all the volunteers and staff who helped.'

'We had hundreds of wooden specimen boxes full of moths and butterflies. Some were specimens from field trips made decades earlier, but most were specimens donated by amateur lepidopterists.'

There will be no wooden storage in the new collections building because the wood could attract pests, especially in the absence of naphthalene as a chemical deterrent. But moving the specimens into metal drawers does more than help protect them from pest insects.

'The other downside of wooden specimen boxes is that the specimens are much less accessible for use in research than the specimens in our standard drawers with glass lids,' Thekla says.

Thekla shows me some drawers, post transfer. They are laid out beautifully. There's a drawer of exotic Sphingidae, or hawkmoth family; 'exotic' meaning from outside Australia. There appear to be many different species here, with a large range of sizes and patterns. Some are more orange, others more grey. Some have eyespots on their lower wings.

Thekla points out that the task of curating these specimens still lies ahead. This will involve checking species identifications and pinning like with like across the more than 22,000 drawers in the insect collection, following the most current classification to keep the collection up to date. It's tasks like these that keep curators busy indefinitely.

A decathlon of beetles

There are many rivalries among people who work on different taxonomic groups, usually of the tongue-in-cheek kind. Why bother researching orchids? Don't we already know enough about birds? (In case you're wondering, the answers are conservation and no.)

Beetle researchers seem to have a quiet confidence that they've made the right choice. Around a quarter of known animal species on Earth are beetles. They are hyperdiverse, ecologically important and possibly the most endearing of all insects.

Dr Yun 'Living' Li is a postdoctoral researcher at the Australian National Insect Collection. He takes a break from prepping beetles for their big move to the new collections building to meet me in the café at work, where we talk about the beetles he studied during his PhD. He says the big question is: why have beetles been so successful?

'How did their enormous diversity arise?' Living says. 'The secrets behind this lie in natural history collections.'

Living studied the Tenebrionidae, or darkling beetles, a family with more than 30,000 species. Their name means 'dark loving', which hints at a group of small, brown critters that are active only at night. That couldn't be more wrong.

Living used 300 specimens in the collection to compare the DNA sequences, ecological roles and body shape morphology of darkling beetles. It led to some surprises. He shows me the evolutionary tree of darkling beetles that he reconstructed using DNA sequences.

'We found that darkling beetles underwent a big bang of evolution near the end-Cretaceous mass extinction, when dinosaurs went extinct. They colonised a vast range of terrestrial environments across the planet, evolving rapidly into a spectacular array of body forms that we see today. Our new research suggests ecological adaptation has driven darkling beetle diversity.'

Living's paper based on his PhD research has just been published in a scientific journal. We try to come up with a creative way to promote the research via CSIRO's blog and social media accounts. With the Olympics underway in Paris, we decide to pit darkling beetles against each other in a decathlon. It's an elaborate thought-experiment with no right answers. Halfway through I start to wish we'd chosen the pentathlon, but here goes.

9 – Big moves and quantum leaps

For the 100 m sprint we decide on the Yellow Mealworm Beetle (*Tenebrio molitor*) for its speed. You might have eaten the larvae of this species, which are used as both pet and human food. The adults are black and active at night, like you might expect a 'darkling' to be. But darkling beetles have a huge variety of shapes and lifestyles.

'You can find darkling beetles in your backyard, rainforests, deserts, mountaintops, alpine heaths and more,' Living says.

Long jump is difficult to match, because no living species of darkling beetle can jump. Living suggests the Egyptian Darkling Beetle (*Blaps polychresta*), a species from Africa that has become a pest in semi-arid areas of southern Australia. It can do a headstand, which resembles an attempt to jump, but when it's in this posture it's actually threatening to spray a smelly liquid as a defence mechanism.

Shotput is obviously the domain of dung beetles. But dung beetles are in a in a different family: Scarabaeidae. No other beetle even qualifies for this event, except the dung-rolling weevils we met in Chapter 6.

We award the high jump to the Shiny Darkling Beetle (*Amarygmus morio*). This species comes out at night to feed on lichen and fungi on the surface of tree trunks. When startled, it drops to the ground and plays dead. This is more of a long fall than a high jump, but it will do.

On the topic of jumping, Living says rapid jumps, or quantum evolution, explain how darkling beetles have evolved a wide array of body forms, from the wood-boring, cylindrical form that resembles a candlestick to surface-grazing, hemispherical forms reminiscent of ping-pong balls.

'Quantum evolution occurred frequently across the evolutionary tree of darkling beetles,' he says. 'This allowed these beetles to shift their body forms to rapidly adapt to changing environments, a secret to their ecological success.'

The 400 m race goes to the Fog-basking Darkling Beetle (*Onymacris unguicularis*) for its speed in inhospitable environments. This species lives in the Namib Desert, a coastal desert in southern Africa. In the early morning it does a headstand, which allows microstructures on its body to condense water from fog, and flow the water into its mouth. In the heat of the day it scurries in search of food, its long legs keeping its body clear of the hot sand.

Living tells me darkling beetles originated from a common ancestor that thrived in humid forests around 150 million years ago.

'Arid adaptations arose at least 17 times in darkling beetles, allowing them to survive in some of the harshest environments on Earth,' he says.

The 110 m hurdles is won by the Giant Darkling Beetle (*Cyhaleus imperialis*), a 5 cm long beetle from Australia's wet tropics. It scurries along the ground among leaf litter and over dead logs, just like doing the hurdles. Like the winner of the long jump, it has a powerful chemical defence system that it can spray from its bum.

'Darkling beetles have gained and lost traits like chemical defence systems many times throughout their evolution,' Living says. 'This used to make it difficult to use their morphological traits to tell which species were closely related. Using DNA sequences to estimate relatedness solves that problem.'

For the discus throw we decide to focus on body shape, and select the Hemispherical Darkling Beetle (*Derispia variabilis*). This darkling species is unique in both its body shape and its ecology. It resembles a Ladybird (family Coccinellidae), its shape protecting its legs and antennae from predators while it forages on rocks or wood for algae, lichen and moss.

The Hemispherical Darkling Beetle is closely related to the Cylindrical Darkling Beetle (*Corticeus unicolor*), which has a very different shape. Its long, thin body is adapted to its habitat – the tunnels made by weevils known as ambrosia beetles, which burrow into trees to farm fungi for their larvae.

'We found darkling beetles that are closely related can look very different,' Living says. 'To find out why, we compared darkling beetle body shapes with their ecology. We found evolutionary changes in body forms have been driven by shifts in the environments where darkling beetles live.'

Living suggests the Green Comb-clawed Beetle (*Lepturidea viridis*) for another track and field event – the pole vault. It's another slim-bodied darkling, which climbs thin poles (the skinny stems of plants) to perch on leaves or flowers. It's a pollinator and part of a group nicknamed 'darkling skywalkers'.

'Our research showed that the slim, skywalker body forms have evolved many times in distantly related darkling beetles that can utilise the soft parts of plants,' Living says.

The javelin throw was won by *Onychocerus albitarsis*, a species from Peru that uses its long antennae to sting. It's in the longhorn beetle family, Cerambycidae, which have long antennae, affectionately known as horns. This peculiar species is the only beetle with a venomous sting. It's also the only insect that stings using its antennae. The stinger of bees and wasps is a modified ovipositor, the structure used to lay eggs. Flies bite.

The last event in the decathlon is the 1500 m, won by the strangest-looking darkling in the competition: the Pie-dish Beetle (*Helea monilifera*; Plate 22). This species is an endurance runner that lives in the arid deserts of Australia. It forages at night and can travel a long way in search of food. Its body is a strange shape, like a pie dish. Living once watched one of these beetles defend itself against ants. It tipped over, using its pie dish like a Roman shield.

'This species is part of a group that includes more than 500 species on the Australian continent,' he says. 'Species in this group evolved different body forms quickly at a time when Australia's environment was shifting into a largely arid continent.'

Living and I leave the café, me to craft a communication campaign to publicise the research and Living to return to the insect collection to continue packing. The insect collection houses 6,000 beetle drawers, each containing many specimens. That's a lot of critters to relocate safely.

Behind it all is a deep appreciation of nature and a sense of how important biodiversity is to our lives.

10
A species on life support

Tasmania was once surrounded by majestic seaweed forests made up of a species called Giant Kelp. Growing as tall as 30 m, these forests provided habitats for countless species of sea creatures. But as climate changes and oceans warm, the forests are vanishing.

Why does biodiversity matter?

Biodiversity means the variety of living things. Implicit in the word is that the more variety, the better. This seems to be true on any scale, from an entire ecosystem to the microbiome within our own guts.

Dr Andrew Young is the former director of the National Research Collections Australia. He has a succinct answer to questions about why biodiversity matters.

'Biodiversity is our free, planet-scale life support system,' he says.

Through an intricate network of interactions, biodiversity looks after Earth. It sequesters carbon, cycles nutrients, purifies water, regulates ocean and atmospheric temperatures, turns waste into soil, pollinates plants, suppresses pests, weeds and diseases, and provides food and shelter to sustain itself.

Since farewelling the director role, Andrew has gone back to his roots in plant conservation genetics, returning to the office where he began his career at CSIRO more than 30 years ago. He's painted the office walls white, polished the vinyl floor and kept the old notes on the whiteboard behind the door, which had somehow survived the three decades since he wrote them.

Andrew has picked up the threads of some old work on Wild Radish (*Raphanus raphanistrum*), a weed that causes a trifecta of problems by competing with crops, hosting pests and diseases, and causing health problems for grazing animals. But he also wanted to work on something new, something that would feel as exciting today as the field of conservation genetics felt back in the 1990s. He found it in an ambitious project to use collections to understand biodiversity change on a planetary scale.

The difference in Andrew's work then and now is itself one of scale. In the past, researchers worked with the specimens they could get their hands on. Studying a few dozen or a few hundred specimens could take months of travelling between herbaria or waiting for specimen to be loaned by mail. Now, with collections going online, it's possible to extract data from huge numbers of specimens, potentially tens of millions, without even leaving your house.

The projects to digitise the CSIRO's collections aim not just to make them vastly more accessible to researchers worldwide, but also to enable new kinds of science in the future.

'Lots of people are digitising collections,' Andrew tells me, 'The exciting thing now is to understand how we can use these new digital assets to know more about our biodiversity so we can make better environmental decisions.'

Andrew has teamed up with long-term collaborator Dr Pete Thrall, head of the collections' digital team, and postdoctoral researcher Dr Owen Forbes, who specialises in data science, to look at whether flowering time in acacias has changed in recent decades due to climate change.

There are around 1,000 different species of acacias in Australia, including Golden Wattle (*Acacia pycnantha*), the inspiration for the country's sporting colours, green and gold. There are plenty of acacia specimens in the Australian National Herbarium, collected from all over the country, throughout different seasons and across decades. The 800,000+ herbarium sheets – pressed specimens mounted on roughly A3 sized paper that we got a glimpse of in Chapter 1 – have already been imaged in high resolution, letting Andrew and Owen compare the flowers of acacia specimens collected at different times of year.

'To start with, we looked at plus and minus,' Andrew says, meaning whether a specimen has flowers and thus whether the plant was in bloom on the date it was collected.

'This has showed us that flowering time is under strong environmental control. We can now use this information to model and predict what will likely happen in the future as climate continues to change, and what that means for the ecological threat to different species.'

'Now we are going to train AI to look at the finer details. Does the specimen have buds or flowers? Are there seed pods? If so, are they juvenile, mature or empty? Taking the AI approach means we can also look for evidence of browsing and disease damage on specimens that will give us important clues about how herbivore and pathogen distributions are changing along with climate.'

'Now imagine doing this using every herbaria on the planet. We could look at flowering time in all flowering plant species, using specimens that date back hundreds of years.

It would show us how whole communities like tropical forests or arid grasslands are responding to environmental change, including climate change.'

Andrew pauses to emphasise that flowering time is under genetic control. This project involves working with images, not DNA, but it does share similarities with conservation genetics. And it's driven by collections.

'There are three big values that collections give us,' Andrew says, listing them on his fingers. 'One is taxonomic breadth. Collections aim to represent every species there is. Next is geographic range. Specimens have been collected all over the world. The last is temporal depth. The world's herbaria have specimens dating back hundreds of years.'

'No other kind of data can do this. Satellite data has poor taxonomic resolution because it's hard to identify plants to species level from space. Monitoring plots, where you peg out an area and observe all the plants, give you the species level data, but they are limited in their geographical range. eDNA studies have no temporal depth because they are so new.'

'Collections can do it all. This new work is just as interesting and useful as getting into conservation genetics was 35 years ago and I think it's going to have at least as much impact for how we manage the planet.'

Beauty and the bees

Biodiversity is beautiful in the eyes of its beholders, fish of the deep sea notwithstanding. Beauty is how plants attract pollinators and animals attract mates. Appreciation of beauty is how human psychology reminds us to do more of what is good for us: be in nature, value nature, look after nature. We need nature to survive. If we had to build any of nature's processes from scratch, it would be an engineering project as complex as establishing a colony on Mars and probably just as expensive.

Biodiversity also supports industries. Beautiful natural places are the foundation of the tourism industry. Fisheries rely on wild populations of wild species. Pollination – the movement of pollen from male flower parts to female flower parts – is essential for agriculture and insects are vital for getting the job done.

It's not well known that vanilla, which comes from the fruit pods of an orchid, needs to be pollinated by people. The plant's specialist insect pollinators live in its native range in Mexico. Grown elsewhere, in the absence of these insects, pollinating vanilla flowers is a delicate task that needs to be completed by hand.

The European Honey Bee (*Apis mellifera*) is the usual hero in pollination stories. It's a domesticated generalist pollinator that has been introduced around the world. In some places, it is even trucked around to do the job, such as in California where hives are brought in to pollinate almond orchards during the flowering season. But the pollination story is much bigger than bees. In Australia, fruit bats pollinate eucalypts, beetles and thrips pollinate cycads, and moths pollinate cheese trees (*Glochidion* species) to name just a handful. And then there are our native bees.

Some native bees are as just as effective at pollination as the European Honey Bee. Australia has 1,653 described species of native bees at the time of writing (Plate 23). Some are the solitary kind that might inhabit bee hotels, those handmade homes of drilled timber and hollow straws that people keep in their gardens. Blue-banded bees are a group of species that buzz pollinate, meaning they vibrate their bodies to help pick up pollen from the flowers they visit. Most native bees don't make honey, but the Sugarbag Bee (*Tetragonula carbonaria*) does. It's also stingless and pollinates some native orchids in the genus *Dendrobium*.

Dr Juanita Rodriguez is a hymenopterist – an expert on bees, wasps and ants – at the Australian National Insect Collection. As well as studying the spider wasps we met in Chapter 4, she studies Australian native bees.

'Native bees play are really important pollinators in both nature and agriculture,' Juanita says. 'Unfortunately, we lack understanding of their biodiversity and ecological roles. There are more than 500 species awaiting scientific classification.'

Juanita is hoping to use the insect collection to finish the job. The collection has 55,000 specimens of native bees collected over many decades from around Australia. Many of these specimens belong to undescribed species. The purpose of classifying them goes way beyond the desire to understand their relationships.

'Our research will help reveal which native bees could act as alternative pollinators in agriculture,' Juanita says. 'This could benefit growers by extending growing seasons for fruits and vegetables or it might provide pollination security for speciality crops.'

Juanita is using a combination of DNA-based and morphological methods to sort out the taxonomy of Australian bees. She's also helping establish a DNA reference library for Australian bee species, which will mean species can be identified from traces of their DNA left in the environment.

10 – A species on life support

Some of the bee specimens Juanita is working with still have pollen stuck to their bodies, decades after they were collected. This pollen can reveal what was flowering in the place and at the time the bee was collected.

Dr Liz Milla and Dr Francisco Encinas-Viso research pollination at CSIRO. They use a technique called DNA metabarcoding to identify plant species from DNA in pollen samples. They can use pollen attached to old bee specimens, traces of pollen in jars of honey, or even pollen collected by bees that they deploy as field scientists, thus using hives to detect flowering plants in the nearby area.

'Honey bees are experts at surveying flowering plants,' Liz says.

She shows me a video of bees passing through a pollen trap as they enter their hive. The trap is bright yellow, an attractive colour to bees.

'Pollen traps are commonly used by beekeepers to collect pollen,' she says. 'They are plastic boxes placed over the entrance to the hive. To enter, bees are forced to pass through little holes that are just big enough for their bodies, but not for the pollen collected on their hind legs, which is scraped off as they force themselves through. The pollen falls into a container at the bottom, which can fill up on busy days.'

'Pollen traps cannot be used for too long as they restrict the pollen resources reaching the hive. This is one of the reasons we are trying to develop less invasive ways to collect floral DNA from bees. So far, we've had good success with forensic swabs. They are much easier to collect and better for the bees.'

Liz and Francisco set up a study at a nature reserve in Canberra to test bees from commercial hives against botanists in a survey of flowering plants. The bees detected more flowering plants than the botanists, but the botanists were better at identifying plants all the way to species level. Altogether, the plants detected by bees and botanists only overlapped by about 25 per cent.

'Bees and botanists should definitely work together to get the best results,' Liz says.

This makes sense, as botanists can recognise plants that aren't in flower on the day of a survey and bees have an edge in places that are difficult for people to access.

'In remote areas, bees could be used as sentinels to provide early warnings of invasive weeds or to confirm the presence of rare, endangered plants,' Liz says.

To demonstrate this, Francisco and Liz set up study plots in the fragile and threatened alpine meadows in Kosciuszko National Park, New South Wales to find out who pollinates what. They analysed the pollen collected by different insect pollinators, including bees, beetles and flies. The result was a complex network of pollination interactions in the alpine meadows.

'The exciting thing about this work is the potential to use pollinating insects to track fine-scale changes in ecosystems over time,' Francisco says.

Like Andrew's work with acacia specimens, time and again research comes back to that same threat, change, and the same scientific problem, how to measure it.

Tiny moth pollinators

Any animal that visits a flower for food or shelter can act as a pollinator, even a tiny moth.

Boronias have intricate relationships with a family of tiny, silvery moths called Heliozelidae. They are active during the day, like butterflies. Their larvae feed on developing boronia seeds and, when mature, drop to the ground and pupate in the soil. Adults emerge in late winter and spring when the boronias are in flower.

Liz says many heliozelid species seem to simply feed on boronias. But some heliozelid and boronia species have evolved an intricate relationship in which the moth also pollinates the boronia. In some cases, this seems to have resulted in a mutually obligate pollination relationship.

'This means the moth is the only species that can pollinate the plant, and the plant is the only species that can feed and shelter the moth and its larvae,' Liz says.

'I think this is the case for *Boronia megastigma* and its moth, which I studied during my PhD. The adult moths pollinate the flowers after laying eggs within them, so their larvae will be able to feed on some, but not all, of the seeds.'

'Female moths have a scaly abdominal cleft that they use to collect and transfer pollen between plants. We think this structure may have coevolved together with changes to the flowers of *B. megastigma* and other species of boronia that have very distinctive flowers,' she says.

Liz and Francisco are now studying populations of *B. megastigma*, in south-west Western Australia.

'We want to find out whether the moths, which are very tiny, can travel between remnant patches of boronias. This is important for conservation because the plants can't set seed without their moths,' she says.

Boronias are part of the citrus family, Rutaceae. Some boronia species have very small distributions, growing in isolated patches in rocky areas or in native vegetation remnants on farms and along roadsides. Others are quite widespread.

Dr Marco Duretto is a botanist at the National Herbarium of New South Wales. He says Australia is a centre of diversity for the citrus family, including boronias. Of more than 130 described species of the *Boronia* genus, a few are from New Caledonia and the rest are endemic to Australia.

'Boronias occur from Cape York to southern Tasmania, with centres of diversity in south-eastern Australia, Arnhem Land, the Kimberley and south-west Western Australia,' Marco says.

'The flowers of boronia species that we think may have heliozelid pollinators appear to have evolved wacky morphology. They have features like petals that are unusual colours or different colours on either side or that barely open; giant, black, sterile stamens; large bulbous stigmas; beautiful perfumes; and other floral features to direct moths to lay their eggs in specific parts of the flower.'

'Other boronias associated with more generalised moths and other pollinators have simple flowers.'

Interestingly, Marco's recent evolutionary studies of boronias using DNA data showed that boronias with these interesting floral features aren't closely related to each other.

'It's a kind of parallel evolution. We think some of these floral features evolved several times, quite recently, in species with specialised moth pollinators.'

Both Liz and Marco note that the boronias species they think have heliozelid pollinators are species that produce beautiful floral fragrances. They assume this is to attract their pollinators. The Rose Boronia (*B. serrulata*), which grows in Sydney is one. And the fragrance of Brown Boronia (*B. megastigma*), which Liz worked on during her PhD, is well known. The species is cultivated from cuttings and grown on farms to isolate chemicals from its flowers for use in fragrances.

With some boronias and moths so intricately linked, what does this mean for conservation? As local climates change, might plants alter their flowering times and finish flowering before the adult moths emerge to pollinate them?

Pollination relationships could also be affected by greater impacts of fire on ecosystems. Plants can regrow after fires. However, if their moth pollinators don't survive, they could find themselves growing in pollinator sterile areas, unable to set seed.

A significant skink

Traditional Indigenous knowledge and Western science are coming together to monitor and protect the Tjakura (Great Desert Skink, *Liopholis kintorei*). The Tjakura is a threatened species of cultural significance to Anangu, and Traditional Owners of Uluru-Kata Tjuta National Park.

Dr David Thuo studied cheetahs in Kenya before moving to Australia to take up a postdoctoral research position at the Australian National Wildlife Collection. Thuo has been working alongside rangers from Uluru-Kata Tjuta National Park and the Central Land Councils Tjakura Rangers to combine insights from his eDNA monitoring with the traditional knowledge Anangu use to locate burrows and track the species.

'Anangu traditional knowledge is used to locate Tjakura burrows and estimate occupancy,' Thuo says. 'I used eDNA collected from soil and Tjakura scats to confirm the number of individuals and analyse what they eat.'

'Extracting eDNA from soil and scats is a way of studying Tjakura without disturbing them.'

A deceased Tjaku_ra found during a survey was donated to the wildlife collection for future genetic studies. Thuo's next step is to use Erin and Clare's techniques for working with DNA from museum specimens (see Chapter 8) to compare the genetic differences between specimens collected in the 1960s, currently held in a museum in Darwin, with the present-day population.

The Tjaku_ra project will continue through long-term monitoring of the burrows to observe how the population is faring, and continue to combine eDNA technology with traditional Indigenous knowledge to monitor and protect the Tjaku_ra.

Saving an ocean forest

In the winter of 2023, Dr Anusuya (Sui) Willis, director of the Australian National Algae Culture Collection in Hobart, alerts me to a new project. Her laboratory is starting to work on a species of seaweed called Giant Kelp (*Macrocystis pyrifera*). The work is supported by Google, which is deeply involved in the science, and they would like to run an environmental awareness campaign about the decline of Giant Kelp forests.

Giant Kelp is a story about hope.

Giant Kelp is the largest of the more than 100 species of kelp. It grows in cold, shallow seas around the world, including South Africa, California and Tasmania.

Dr Jakop Schwoerbel is a postdoctoral researcher and algae taxonomist at the algae collection.

'Giant Kelp has a holdfast that attaches to hard substrate such as rocks on the sea floor,' Jakop says. 'Long stipes support leaf-like blades that each have a gas bladder at their base to keep the kelp afloat. An individual can reach well over 50 m

10 – A species on life support

tall, but in the seas around Tasmania they can grow to around 30 m tall.'

Growing together in the ocean, Giant Kelp forms forests like trees on land. These marine forests support an entire ecosystem, including young fish sheltering in nurseries, marine plants and other seaweeds, sea lions, marine invertebrates like octopuses, sea stars and jellyfish, and even insects like kelp flies, which recycle giant kelp that has detached and washed ashore at the end of its life.

Towering forests of Giant Kelp once surrounded Tasmania. As the climate continues to change and ocean temperatures increase, the East Australian Current has moved further south, bringing warm, nutrient-poor waters to Tasmania's seas. It's also brough a migrant from New South Wales, the Long-spined Sea Urchin (*Centrostephanus rodgersii*), which grazes on Giant Kelp. The kelp is very sensitive to warmer water and is suffering from the loss of nutrients and grazing.

With only deep ocean between Tasmania and Antarctica, Giant Kelp has no shallow, cool seas to retreat to. Without these seaweed giants, the whole Giant Kelp forest ecosystem shrinks. Giant Kelp Marine Forests of South-East Australia are now nationally listed as an endangered marine community type.

After my conversation with Sui, I start working with Google, University of Tasmania's Institute for Marine and Antarctic Studies (IMAS), The Nature Conservancy (TNC) and other partners on a campaign to raise public awareness of the plight of Giant Kelp and the science to help restore the forests.

I visit Sui's team in Hobart to photograph their work. They grow Giant Kelp in the laboratory by working with the microscopic stage of its life cycle. They have some tiny babies that are barely a centimetre across, growing in sea water in culture plates that have multiple wells, like tiny petri dishes (Plate 24).

Giant Kelp alternates between two phases during its life cycle, a giant sporophyte and a microscopic gametophyte. When Giant Kelp reaches maturity or around 5 m tall, it begins to release microscopic spores from specialised reproductive blades, called sporophylls, that grow near its holdfast. The spores settle nearby on the sea floor and grow into microscopic female or male gametophytes that produce eggs or sperm. Once fertilised, the eggs grow from the female gametophytes, forming baby sporophytes that begin to grow into giants. At its peak, Giant Kelp can grow 50 cm in a single day.

Dr Hugo Scharfenstein is a postdoctoral researcher at the algae collection. He collected some of this giant seaweed species from remnant populations off the eastern coast of Victoria, taking tiny samples to obtain reproductive cells for breeding Giant Kelp and tissue for DNA studies to help conserve the species.

'We chartered a boat from a local diver, who knew the locations of the kelp,' Hugo says. 'The environment it was growing in varied from small patches in water only 1–2 m deep, which you could see on the surface, to a larger, quite healthy patch where the individuals were 4–6 m tall in ~8 m of water. This was a rocky channel where Giant Kelp was growing with other forms of kelp, a few fish and one tiny shark.'

Next, I visit the original Giant Kelp gametophyte biobank at IMAS, just down the road from our algae collection, to take more photos. There are dozens of round flasks of sea water, bubbling as air is pumped though plastic tubes. The entire set-up glows in red light.

Under red light at 4°C, gametophytes will remain in a kind of stasis and so can be stored long-term as a biobank, like a seed bank, to protect the species. Warmed to 12–15°C, gametophytes will grow and produce more cells for study. Blue light will trigger them to release eggs and sperm. The lighting doesn't need to be any more high-tech than a piece of red or blue cellophane wrapped around a test tube.

Next to the biobank, IMAS has a couple of small aquarium tanks where hundreds of Giant Kelp babies are growing on twine wrapped around PVC pipe, waiting to be replanted in the ocean on the long lines of twine. These babies are then planted onto natural reefs as part of large-scale restoration trials.

The team hopes to boost the resilience of their restoration efforts by identifying and propagating Giant Kelp that is naturally more tolerant of warmer water. Sui's laboratory is supporting this by diving into the DNA of Giant Kelp to discover why some strains are more tolerant of warmer sea temperatures (Plate 25).

'To help select the right strains, we are using techniques like whole genome sequencing,' she says. 'That involves looking at the entire genome to pinpoint particular genetic differences associated with warm tolerance.'

'We have some initial evidence that there is a genetic basis for warm tolerance. We're also studying the genetic structure of the remnant populations of Giant Kelp in eastern Tasmania. It's important to maintain all of the other genetic diversity as well as selecting for thermal tolerance.'

'We hope our work to understand Giant Kelp populations will enable us to select warm-tolerant strains for replanting. This will give Tasmania's Giant Kelp forests the best chance to recover and flourish into the future.'

I pay a visit to the Tasmanian Herbarium to see their specimens of Giant Kelp. Like at our herbarium in Canberra, there are rows of mobile shelving storing stacks of plant specimens attached to sheets of archival paper approximately A3 in size. It takes us a moment or two to recall the species name: *Macrocystis pyrifera*. Then the specimens are found quickly and I'm able to observe and photograph some Giant Kelp collected in the 1950s, just before the forests began to decline. The leaf-like blades are a soft brown colour and slightly transparent, with tiny spikes along their edges.

Seaweeds have many superficial similarities to plants. Holdfasts, like roots, anchor the seaweed to the sea floor, stipes function like stems, blades are like leaves and they photosynthesise. The three different groups of seaweeds – browns, reds, greens – are named for the pigments they use to absorb different wavelengths of sunlight at different ocean depths.

'The browns, reds and greens are only distantly related,' Jakop says. 'Brown seaweeds like Giant Kelp contain chlorophyll and carotenoid pigments. The carotenoid pigment fucoxanthin overshadows the green chlorophyll and gives Giant Kelp its golden brown colour.'

The specimens show that the blades of Giant Kelp grow from the same side of the stipe. The floats at the base of each blade look like those of any seaweeds washed up on the beach.

In February 2024, all the project partners meet in Hobart to introduce a group of journalists and online content creators to the restoration project.

We are joined by Emma Robertson, a Palawa artist, who gives a Welcome to Country and shares details of how important Giant Kelp is to her People. Giant Kelp protected Palawa women from sharks as they swam and fished. Kelp was often used to create water carriers for collecting and moving fresh water. Emma makes bowls and necklaces from Bull Kelp (*Durvillea potatorum*) and has brought beautiful examples to show us.

The group of journalists and content creators travels by boat to visit a remnant population of Giant Kelp not far from Hobart. From the surface of the sea the canopy of a Giant Kelp forest is visible, its long blades curling and swaying. Part of Google's work involves partnering with geospatial experts at NGIS to map the forests from space using satellite imagery to locate and analyse remnant populations. Google is also working closely with Sui's team to provide AI tools to understand the complex genetics of thermal tolerance.

By the end of the media tour, there's a strong sense of shared purpose and optimism about rewilding Tasmania's Giant Kelp forests. But as David Attenborough said in his 2020 documentary *A Life on Our Planet*, it's time to rewild the world.

I titled this chapter 'A species on life support'. That species is us. During the course of our lives, our moments big and small, our troubles insurmountable and trivial, nature is what matters. Nature is keeping us alive. Nature needs us to return the favour. Collections science is key to achieving this.

GLOSSARY

Algae – a mixed group of (mostly) photosynthetic organisms that covers seaweeds and microorganisms such as cyanobacteria, dinoflagellates and diatoms and a few others

Biodiversity – the variety of living things

Biosecurity – measures to prevent access or spread of invasive species or disease agents such as viruses

Chondrichthyes – a class that includes Elasmobranchii (sharks, rays, skates and sawfish) and Holocephali (chimaeras)

Chromosomes – packages of DNA inside a cell

Class – a taxonomic rank, above Order

Coleoptera – beetles, an order of insects

Common name – a simplified name, sometimes given to a species at the time of scientific description, e.g. Painted Hornshark for *Heterodontus marshallae*

Cryptogams – a group that includes non-vascular plants, such as ferns and mosses, and other organisms, such as lichens, fungi and algae

Diptera – flies, an order of insects

DNA – a biological molecule made up of four different subunits that stores life's genetic code

eDNA (environmental DNA) – traces of DNA left behind by plants, animals, fungi, algae and microbes in the environment, including in air, water and soil

Elasmobranchii – see Chondrichthyes

Entomology – the study of insects

Family – a taxonomic rank above Genus and below Order

Genome – all of the DNA of an individual

Genus (plural genera) – a taxonomic rank above species and below Family

Glossary

Holotype – the specimen chosen to anchor the scientific name of a species or subspecies

Hymenoptera – bees, wasps and ants, an order of insects

Lepidoptera – moths and butterflies, an order of insects

Lineage – descent of a species from an ancestral species

Morphology – the physical features of an organism

Natural history collection – a set of specimens preserved and curated for scientific research

Nematode – a roundworm, often microscopic, which may be free-living or parasitic

Order – a taxonomic rank above Family and Genus

Oviparous – egg-laying

Paper – research by a scientist or a team of scientists, published in a scientific journal after formal peer review

Parasite – an organism that lives in symbiosis with a host organism and benefits at the host's expense

Phylogeny – the evolutionary history and pattern of relationships among species, often represented graphically as a phylogenetic tree

Pollination – movement of pollen from the male to female parts of a flower, necessary for fertilisation so the plant can produce seeds

Population genomics – a way of comparing DNA variation between different populations of a species

Species name – a scientific name given when a species is formally described by a taxonomist, e.g. *Heterodontus marshallae* (Painted Hornshark), often then abbreviated, e.g. *H. marshallae*

Specimen – a part or whole plant, animal or other lifeform stored in a natural history collection for scientific research

Taxonomy – the science of describing and naming species

Plates

Plate 1 A specimen of the waterlily *Nymphaea violacea* in the Australian National Herbarium. Photo by Australian National Herbarium.

Plate 2 An orchid stored in ethanol at the Australian National Herbarium. Photo by Andrea Wild, CSIRO.

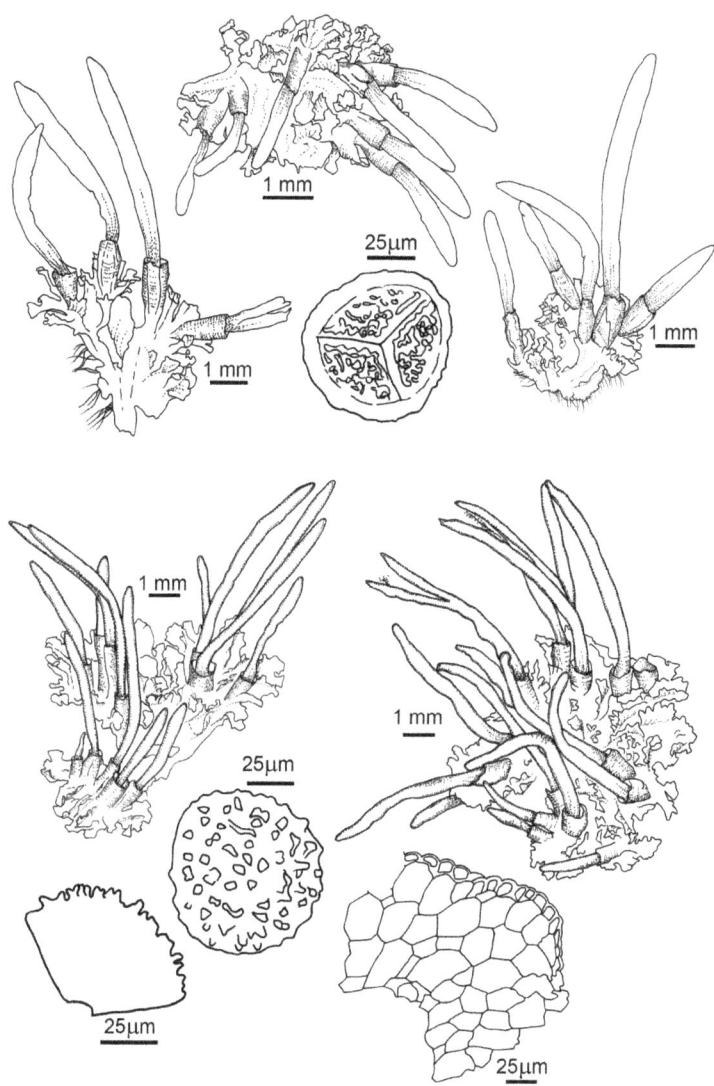

Plate 3 Line drawings used in the species description of the hornworts *Anthoceros apocynon* and *A. wellmanii*. Line drawings provided by D. Christine Cargill, Australian National Herbarium.

Wild Collections

Plate 4 Dr Katharina Nargar with the orchid *Dendrobium nindii* at the Australian Tropical Herbarium in Cairns. Self portrait.

Plate 5 The Painted Hornshark (*Heterodontus marshallae*). Photo by CSIRO.

Plate 6 Deadpool fly (*Humorolethalis sergius*) has markings on its back that resemble Deadpool's mask. Photo by CSIRO.

Plate 7 A specimen of the spider wasp *Ctenostegus immitis* in the Australian National Insect Collection. Photo by Olivia Evangelista, CSIRO.

Plate 8 Dr Francisco Encinas-Viso (left) and Dr Juanita Rodriguez (right) using nets to collect tiny moths on boronias in Western Australia. Photo by Liz Milla, CSIRO.

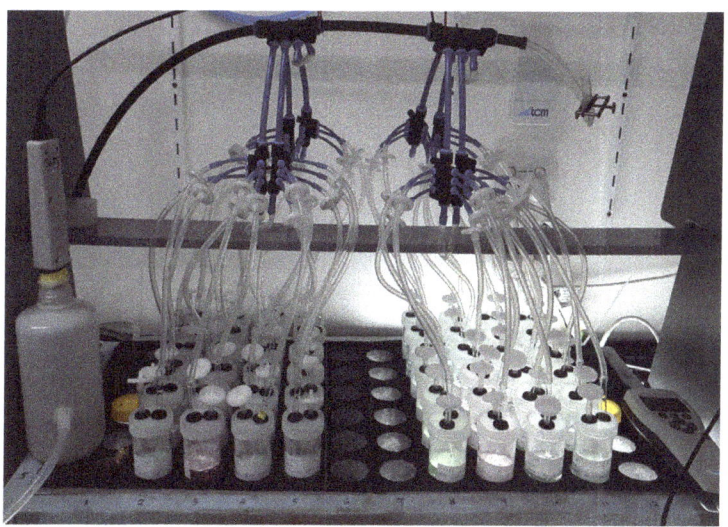

Plate 9 Microalgae strains being tested for CO_2 uptake for use in carbon storage, at the Australian National Algae Culture Collection. Photo by Anusuya Willis, CSIRO.

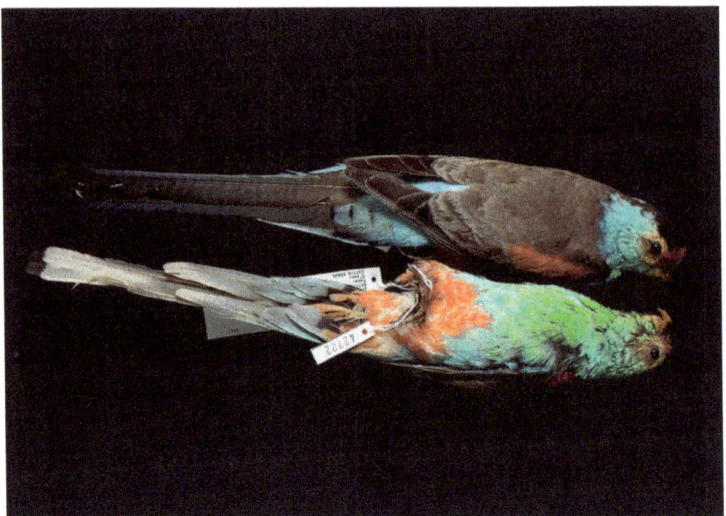

Plate 10 The pair of Paradise Parrots in the Australian National Wildlife Collection. Photo by Gordon Gullock.

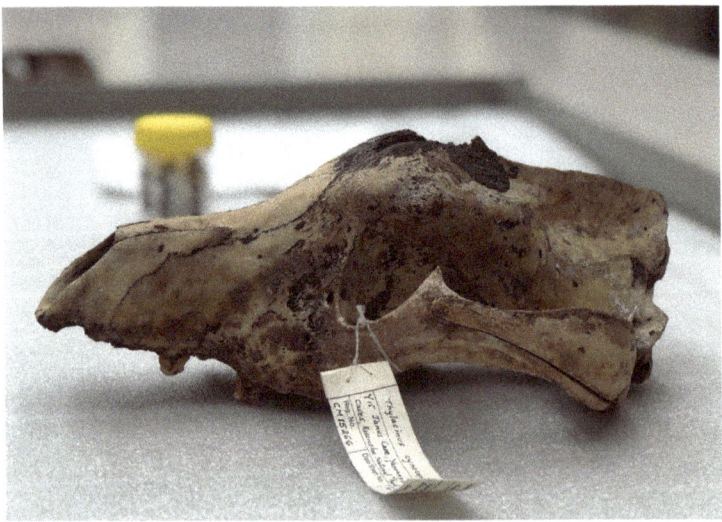

Plate 11 The skull of a Thylacine in the Australian National Wildlife Collection. Photo by Andrea Wild, CSIRO.

Plate 12 The fossil sawfly from McGraths Flat. Photo by Michael Frese.

Plate 13 Painting of the Lost Shark (*Carcharhinus obsoletus*). © Dr Lindsay Marshall (stickfigurefish.com.au).

Plate 14 A weevil rolling dung at Undara Volcanic National Park in north Queensland. Photo by Hermes Escalona, CSIRO.

Plate 15 Helen O'Neill holding a shark egg case at the Australian National Fish Collection. Photo by Carlie Devine, CSIRO.

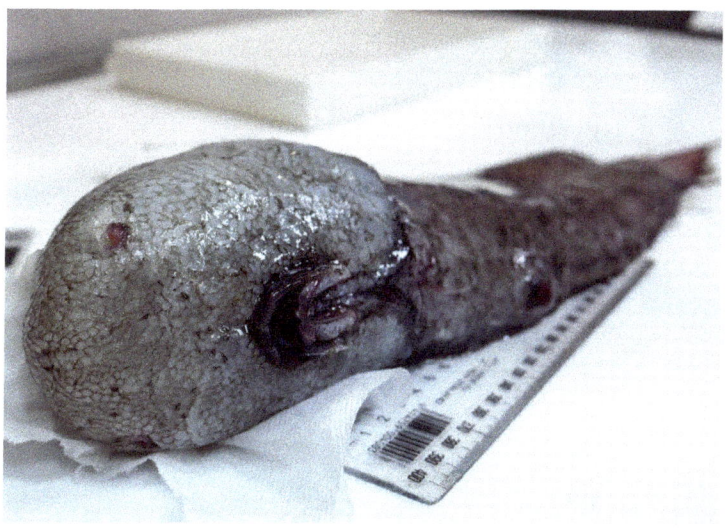

Plate 16 Top: A specimen of a Faceless Cusk (*Typhlonus nasus*). Photo by Asher Flatt, CSIRO and Museums Victoria. Bottom: A Deepsea Lizardfish (*Bathysaurus ferox*) collected from the abyss off Australia's east coast. Photo by Rob Zugaro, Museums Victoria.

Wild Collections

Plate 17 Specimens of a Swift Parrot and a glider in the Australian National Wildlife Collection. Photo by Gordon Gullock.

Plate 18 Daisy Fleabane growing in the wild. Photo by Alexander Schmidt-Lebuhn, CSIRO.

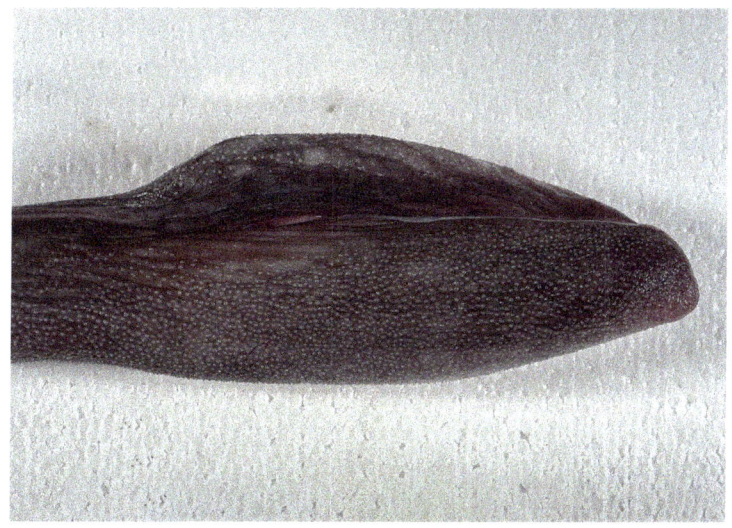

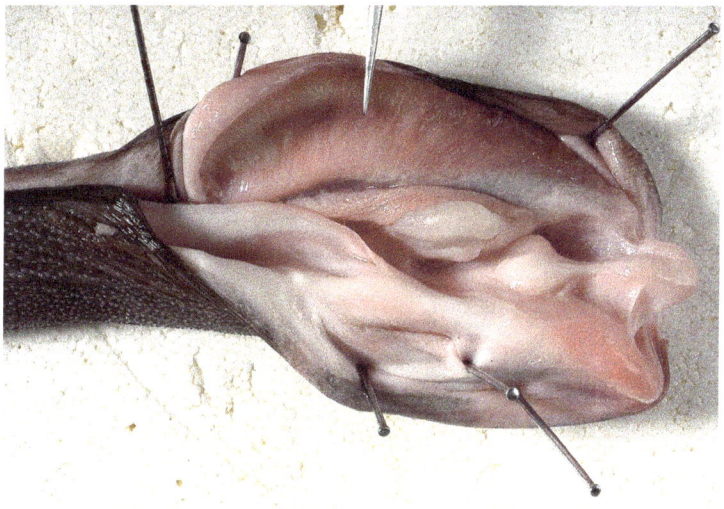

Plate 19 Clasper of a Melbourne Skate (*Spiniraja whitleyi*) not dilated (top) and dilated and pinned to show its structure (bottom). Photos by CSIRO.

Plate 20 The new national collections building, Diversity, in Canberra. Photo by Andrea Wild, CSIRO.

Plate 21 Staff at the Australian National Wildlife Collection packing the egg collection ahead of the move to the new collections building. Photo by CSIRO.

Plates

Plate 22 The darkling beetle *Helea monilifera*. Photo by Yun 'Living' Li, CSIRO.

Plate 23 A specimen of the native bee *Amegilla (Asaropoda) bombiformis* in the Australian National Insect Collection. Photo by David Yuan, CSIRO.

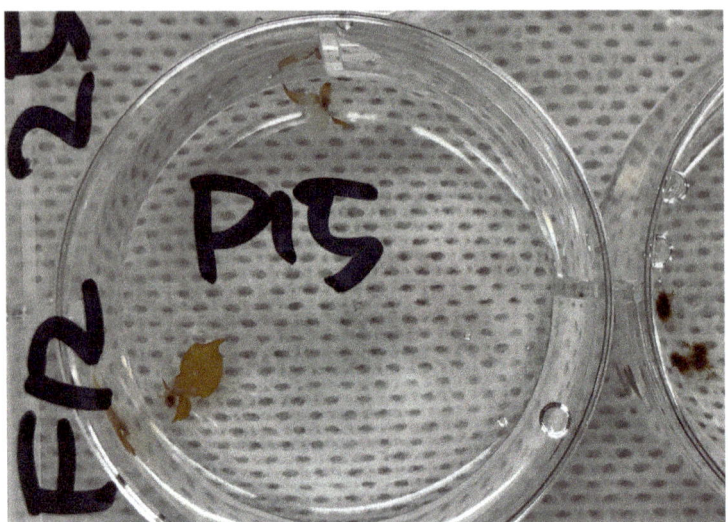

Plate 24 Baby Giant Kelp growing in sea water in the Australian National Algae Culture Collection. Photo by Andrea Wild, CSIRO.

Plate 25 Dr Anusuya (Sui) Willis (left) and Dr Cintia Iha (right) with a piece of Giant Kelp at the Australian National Algae Culture Collection. Photo by Matt Marrison, CSIRO.

INDEX

Page numbers in **bold** refer to illustrations.

Acacia tea 118
acacias 117–18, 145
Acanthobothrium 38
AI, in research 53–7, 110, 145
AI models 105, 106
algae 49, 53–8, 154–9
  birdbath 55
amber 85–6, 127
Amegilla (Asaropoda) bombiformis **177**
Anastatus 16
Anonychomyrma inclinata 34
ants 34, 42
Appleyard, Sharon 102
apps, for species identification 48, 105, 106
Ascidian, Stalked 101
Attenborough, David 69, 159
Australian National Algae Culture Collection 53, 54–5, **169**, **178**
Australian National Fish Collection 25–7, 82, 113–15, **172**
Australian National Herbarium 3–6, 10, 11, 12, 13, 103, 133, 145
Australian National Insect Collection 43, 67, 78–9, 85, 86, 95–7, 106, 110, 121, 127, 130–1, 133–7, 142, 148

Australian National Wildlife Collection 22, 31, 63, 64, 71, 120, 131–3, **176**
Australian Tree Seed Centre 14, 117
Australian Tropical Herbarium 35

bananas 9
BANG mix 10
Bath, Chris 35
Baudin expedition 40–1
Bayless, Keith 27–9, 65–7
Bee, Wallace's Giant 67
bees 30, 42, 67, 111, 147–50, **177**
  native 147–8, **177**
Beetle
  Green Comb-clawed 141
  Pie-dish 142
  Yellow Mealworm 139
beetles 71, 77, 78–9, 93–4, 137–42
  ambrosia 84–5, 141
  blister 110–11
  carpet 96
  darkling 138–41, **177**
  dung 77–80, 139
  scarab 77
Bellerochea 55
Bessey, Cindy 123, 124
Bickerstaff, James 83–5

biodiversity 143–4, 146–7
biodiversity surveys 89
bioluminescence 57, 66
bioproducts 53–4
bioprospecting 60
bird skins 23, 32, 41–2
Blobfish 90
Boea resupinata 34–5
Borer, Polyphagous Shot-hole 83–4
Bornemissza, George 78
boronias 150–3
botanical collecting 3–7
Broom, Cape 103
Budd, Alyssa 125
Bush, David 14–15
butterflies 33–4, 110, 135–6
Butterfly, Bulloak Jewel 34

caddisflies 58–9
caffeine 51
Campbell, Bronwyn 59–60
Cargill, Chris 11–12
Catshark, Banded Sand 26, 27
Chen, Stephanie 107–8
chimaeras 114–15
Chough, White-Winged 31
claspers 39, 114, 116, **175**
Clements, Mark 19, 21
Clisa australis 27–9
coatings 58–9
Cockatoo, Sulphur-crested 99
collection digitisation 13, 131–2, 144
cryptids 68–70, 72
cryptogams 11, 13

cusk eel, blind 115
Cusk, Faceless 90, **173**

daisies 103–8
Darkling Beetle
  Cylindrical 141
  Egyptian 139
  Fog-basking 140
  Giant 140
  Hemispherical 140–1
  Shiny 139
darkling beetles 138–41, **177**
Deagle, Bruce 17–18
Dendrobium 148, **166**
Devil Ray, Giant 25
devil rays 24–5
Devloo-Delva, Floriaan 115–16
diatoms 55–7
digitisation of collections 13, 131–2, 144
dinosaurs 125–6
Diversity (building) 8, 129–37, **176**
DNA, environmental 70, 121–4, 146, 153–4
DNA extraction 9, 22–4, 41–2, 72, 118–21, 126–7
DNA metabarcoding 149
DNA sequencing 24, 70, 86, 100, 107–8, 121, 125–7, 157
Dragon
  Central Bearded 121
  Eastern Water 119
dragons 83
Drummer, Brassy 35
Dunaliella tertiolecta 55
dung beetles 77–80, 139

Duretto, Marco 152
Dyne, Geoff 113

earthworms 113
eDNA 70, 121–4, 146, 153–4
Edwards, Ted 62
egg cases, shark 80–3, **172**
elephant birds 72–4
Elix, Jack 52
Encinas-Viso, Francisco 149–51, **169**
Environomics Future Science Platform 59
ethanol 7, 119–20
ethanol vault 132–3
eucalypts 7, 13, 36
Eucalyptus platyphylla 7
eyes, fossilised 58–9

Fireweed 106–7
fish, and eDNA 123–4
Fleabane
 Daisy 107–8, **174**
 Flaxleaf 107–8
flies 27–30, 35, 36, 65–7
 bot 30
 robber 43
 skipper 66
 soldier 35, 42
flowering time 145–6
Fly
 Bush 27, 78
 Deadpool 43, **167**
 House 27
Forbes, Owen 144–5
formaldehyde 8, 10
formalin 25, 71, 118–19

fossils 58, 68, 85
 amber 85–6, 127
 caddisfly 58–9
 jumping spider 86–8
 mite 112
 pollen 68
 sawfly 67, 68, **171**
frass 96
Frese, Michael 58–9, 68, 86–8, 112–13
Frog, Green Tree 36
fungi 11, 20, 49, 59, 66, 84–5, 94, 127

galah 40–2
Galaxias
 Pedder 17
 Swan 17–18
gametophyte biobank 156–7
Gecko, Bynoe's 120–1
geckos 120–1
genitalia
 cusk eels 115
 insect 109–11, 112
 lizardfish 90
 nematode 109–10
 rays 114
 sharks 39–40, 113–14
genome sequences 22–3, 85, 125, 138, 157
Gerber, Livia 120–1
Gesaia csiro 90–1
Girault, A.A. 15, 16–17
glider **174**
Glider, Sugar 100
Graham, Al 26
Gray, Peter 10

Grealy, Alicia 72–4
Great Eggcase Hunt 82
Groundsel, Gravel 107
Gueidan, Cecile 44–5, 49–52
Gum
    Camden White 13–14
    River Red 36

Haff, Tonya 96
Hahn, Erin 118–20
Handfish, Spotted 100–2
Haslea 56
Haycocknema perplexum 3
helminth collection 10
herbarium specimens 5–7
Hodda, Mike 69–70, 94
Holleley, Clare 8, 100, 118–19
holotypes 32, 39, 41–2
Homo sapiens 32
Honey Bee, European 51, 147, 149
Hornshark, Painted 39–40, 80, **166**
hornworts 11–12, **165**
Housefly 27
Humorolethalis sergius 43, **167**
Huston, Dan 35, 94, 109–10

Iha, Cintia 53–4, **178**
Investigator (RV) 26, 38, 89, 113, 115, 123

James, Tasha 117–18
Joseph, Leo 22–4, 40–2, 98–9
jumping spiders 86–8

Kelp
    Bull 159

Giant 143, 154–9, **178**
keys, identification 104, 105
Kitchen Brush of Science 124
Kraken 69
Kyne, Peter 38

Last, Peter 37, 38
Lee, Stan 43
Lepschi, Brendan 4–7, 13, 102–4
Leptanilla voldemort 42
Lessard, Bryan 35
Li, Yun 'Living' 138–42
lichen 44–5, 49–52
lifespan estimation 121, 125
Lily, Ginger 4–5, 7
Linnaeus, Carl 32
Lizardfish, Deepsea 90, **173**
Loch Ness Monster 69–70

Macbeth 3, 11
Magpies 63–4
Malaise trap 28–9, 66
Marshall, Lindsay 39–40, 75
Marvel Universe 43
McCurry, Matthew 58, 88
McGraths Flat, NSW 58, 68, 112
microalgae 54–7, **169**
Milla, Liz 149–52
mites 112
Monarch
    Black-winged 23–4
    Pearly 24
Monkfish 26
mosquitoes 33, 125–6
mosses 11, 12, 13
Moth
    Case-making Clothes 95

Webbing Clothes 95
moths 62–3, 72, 116, 127, 134–7,
    150–3
  clothes 95–7
Moths and Butterflies of
    Australasia (MABA) 134–5
Mound, Laurence 117

Nargar, Katharina 20–1, **166**
National Biodiversity DNA
    Library 124
native bees 147–8, **177**
Naumann, Ian 15–17, 48
Nematode
  Pine Wood 93–4
  root-knot 109–10
nematodes 1–3, 69–70, 93–4,
    109–10, 112
nets **169**

O'Hara, Kate 120–1
O'Neill, Helen 24–5, 38–40,
    80–3, 101, 113–14, **172**
oak moss 50
Ogyris caelestia 33
Onychocerus albitarsis 141
Opaluma rupaul 42
Ophidascaris robertsi 1–2, 10
Orchid
  moth 19
  underground 19–20
orchids 9, 19–21, 147, 148, **164**,
    166

Paramonovius nightking 42
parasitoid wasps 15–17
parlour domes 61

Parrot
  Golden-shouldered 62–3
  Hooded 62–3
  Night 21–3
  Paradise 61–3, 72, **170**
  Swift 99–100, **174**
parrots 21–3, 61–3, 72, 98–100,
    170, **174**
Patterson, Ashara 116–17
peanut worms 113
penicillin 51
Phaeodactylum 57
Phalaenopsis 19
pharmaceuticals 51–2
Piophilosoma 66
plastic waste 59–60
Platypus 35–6
Pleines, Thekla 136–7
Pogonoski, John 90
pollen, fossilised 68
pollen traps 149
pollinators, insect 51, 147–8,
    150–3
polychaete worms 90–1
Power, Haylea 122, 124
preservatives 10, 118–21
Project Urquhart 69, 70
Python, Carpet 1, 2

quinine 51

Radish, Wild 144
rails 32
reproduction
  geckos 120–1
  Giant Kelp 156
  sharks 39–40, 80–3, 114, **172**

thrips 117
Rhizanthella 19–20
Richardson, Barry 86–8
Robertson, Emma 158–9
Robinson, Isabella 43
Rodriguez, Juanita 48, 67–8, 148–9, **169**
RV *Investigator* 26, 38, 89, 113, 115, 123

Sacred Scarab 77
salicin 51
Salvinia molesta 111–12
Sands, Don 62–3, 112
sawfishes 25
sawflies 67–8
sawfly, fossil 67, **171**
Scharfenstein, Hugo 156
Schmidt pain index 47
Schmidt-Lebuhn, Alexander 104–6
Schwoerbel, Jakop 154–5
scientific naming 32–45
sea sparkle 57
Sea Urchin, Long-spined 155
sea water, eDNA in 122–3
Seastar, Northern Pacific 101
seed germination 20, 117–18
seed orchard 13–14
seeds
   Acacia 117–18
   daisy 104–5
   orchid 20
sex determination 83, 115–16
Shark
   Basking 113
   Broadnose Sevengill 26
   Goblin 37

Lost 61, 74–5, **171**
Port Jackson 40, 81
White 115–16
sharks 26–7, 37, 38–40, 74–5, 80–2, 113–16
   egg cases 80–3, **172**
Simaetha 88
Skate, Melbourne **175**
Skink, Great Desert 153–4
slime mould 11
Solanum elaeagnifolium 103–4
species
   Australian 18
   naming 32–45
specimen preservatives 10, 118–20
spider wasps 47–9, **168**
spiders, jumping 86–8
spirit mix 10
spirit vault 8–10
sponges 26–7
Spratt, Dave 2–3
Squid
   Dana Octopus 71, 124
   Giant 71
squids 71–2, 124
Stemonitis flavogenita 11
stingray
   giant 37, 38
   whiptail 38
Stink Bug, Brown Marmorated 16, 105–6
stink bugs 105, 106
study skins 23, 32, 41–2

Taeniophyllum cylindrocentrum 21
Tapeinochilos 10
tapeworms 10, 38

Index

Tasmanian Museum and Gallery (TMAG) 71–2
Tentegia 79–80
This canus 36
Thrall, Peter 144
thrips 117
Thuo, David 153–4
Thylacine 64–5, **170**
Thyreophora cynophila 66
tissue banks 119
Tjakura 153–4
Tragopogon dubius 103
Trapelia pruinosa 44–5
Trapelia rosettiformis 44
traps
  insect 28–9, 66, 95, 96–7
  Malaise 28–9, 66
  moth 95
  pollen 149
  sticky 96–7
trematodes 35
Trout, Brown 17–18
Turco, Federica 110–11

underground orchid 19–20
Urogymnus acanthobothrium 38

vanilla 147
vaults
  ethanol 132–3

spirit 8–10
venom, spider wasp 47–8
vertebrate collections 8–9, 118–19
Viacava, Pietro 64

wasps
  parasitoid 15–16
  spider 47–9, **168**
waterlily **163**
Watson, Ros 55–7
Wattle, Golden 145
weeds 103–8, 111–12, 144
weevils 35–6, 79–80, 83–6, 112, 116–17, 141, **172**
Wells, Alice 58
West, Katrina 123
Whipray, Leopard 26
whiprays 26, 37–8
White, Will 24–5, 37–8, 74–5
wildlife management 125
Willis, Anusuya (Sui) 55–7, 154, 155, 159, **178**
Wilson, Nerida 124

Yeates, David 97, 125–7
Young, Andrew 143–6

Zich, Frank 35
Zimmer, Heidi 19
zoonoses 1–3

www.ingramcontent.com/pod-product-compliance
Lightning Source LLC
Chambersburg PA
CBHW061245230426
43662CB00020B/2429